FROM FIELD TO FORK

FROM FIELD TO FORK

Food Ethics for Everyone

Paul B. Thompson

OXFORD
UNIVERSITY PRESS

OXFORD

UNIVERSITY PRESS

Oxford University Press is a department of the University of
Oxford. It furthers the University's objective of excellence in research,
scholarship, and education by publishing worldwide.

Oxford   New York
Auckland   Cape Town   Dar es Salaam   Hong Kong   Karachi
Kuala Lumpur   Madrid   Melbourne   Mexico City   Nairobi
New Delhi   Shanghai   Taipei   Toronto

With offices in
Argentina   Austria   Brazil   Chile   Czech Republic   France   Greece
Guatemala   Hungary   Italy   Japan   Poland   Portugal   Singapore
South Korea   Switzerland   Thailand   Turkey   Ukraine   Vietnam

Oxford is a registered trademark of Oxford University Press
in the UK and certain other countries.

Published in the United States of America by
Oxford University Press
198 Madison Avenue, New York, NY 10016

Library of Congress Cataloging-in-Publication Data
Thompson, Paul B., 1951–
From field to fork : food ethics for everyone / Paul B. Thompson.
pages cm
Includes bibliographical references.
ISBN 978–0–19–939169–1 (pbk. : alk. paper) — ISBN 978–0–19–939168–4
(hardcover : alk. paper)   1. Agriculture—Moral and ethical aspects.
2. Food supply—Moral and ethical aspects.   3. Food habits—Moral and
ethical aspects.   I. Title.
BJ52.5.T54 2015
178—dc23
2014041251

1 3 5 7 9 8 6 4 2
Printed in the United States of America
on acid-free paper

Dedicated to Eudora Vasquez

CONTENTS

ACKNOWLEDGMENTS

Chapter 7 is a substantially revised and expanded version of "Ethics, Hunger and the Case for Genetically Modified (GM) Crops," in *Ethics, Hunger and Globalization: In Search of Appropriate Policies* (Pinstrup-Andersen and Peter Sandøe, eds., Dordrecht, NL: Springer, 2007, pp. 215–235). All the other material in this book is appearing here for the first time.

The germ for this book began in earnest during my sabbatical at Portland State University (PSU) where I spent a year as a visiting fellow at the Institute for Sustainable Solutions (ISS). I would like to thank Jennifer Allen, ISS director, and my sponsor, David Ervin, as well as ISS's founding director Robert Costanza, all of whom made my stay there possible. Kim Heavener at PSU made it easy, including putting up with a suitcase in her office when I was away. I must also acknowledge the W. K. Kellogg Foundation for their gift to Michigan State University that supports my chair. I have learned too much from many friends and colleagues over the years to begin thanking all the people who have influenced the thinking that went into this book, but I will mention one who passed away more than a decade ago: Glenn L. Johnson was a professor of agricultural economics at Michigan State University when I first started to work in agricultural and food ethics as an assistant professor at Texas A&M University in 1980. Looking back, I am humbled and amazed at the amount of time Glenn invested in making sure that I did not make an idiot of myself too frequently to survive in the world of farmers and faculty of various colleges of agriculture. Glenn was adamant about two things. First, people working professionally in agricultural science and food production had already thought a lot about ethics. Second, despite this, they needed people trained in philosophy to work through the challenges ahead. Glenn was not about to let me forget the first point, even as he spent many hours laying a foundation for me to help with the second.

I did not start life with a deep interest in food and farming, much less food ethics. Like many in my generation I went into philosophy as an environmentalist committed to the idea that reformulating our values was crucial for the survival of our planet. After completing a dissertation on risk assessment and nuclear energy, I found myself at Texas A&M University, where John J. McDermott, former head of the Department of Philosophy, and H. O. Kunkel, dean of agriculture, were attempting to find someone who would teach a new course on ethics and agriculture. I recall sitting in John's backyard with flames from his barbecue blazing ten feet in the air as he regaled me with stories about how food ethics would be the wave of the future. I was (justifiably) skeptical. This was 1981 and there was a robust ethical literature emerging on what was then called "world hunger" and on the use of animals for food. It would be relatively easy to build a course on those topics, but I did not think of myself as having much to contribute from a research and publication standpoint.

As I have recounted in *The Agrarian Vision*, I was lured more deeply into the field by reorienting some risk-related work that came out of my dissertation. Many of the philosophical questions I had investigated in connection to nuclear power seemed to be relevant to the then-nascent techniques for gene transfer in agricultural plants and animals. I eventually became more deeply interested in sustainability and the cultural significance of the European farming tradition. Within a few years I had become friends with Richard Haynes. The W. K. Kellogg Foundation helped launch a series of conferences that evolved into the Agriculture, Food and Human Values Society (I served as the second president), and things were on their way. Not "the wave of the future" perhaps, but there were interesting questions to ponder and plenty of opportunities to squeeze ideas from ethics and the philosophy of science and technology into agricultural science disciplines. My important work was more likely to show up in *Plant Physiology* or the *Journal of Animal Science* than in *Mind* or *Synthese*. I was very happy with that.

With each passing year, I have come to appreciate more and more Glenn's effort to educate me in the ways of food and farming. Food began to become a subject with wider appeal after massively popular books by Eric Schlosser, Marion Nestle, and Michael Pollan. I started thinking less about taking ethics into the work of food system professionals and began to wonder how I could transfer some of the things I had learned about agriculture to a new generation of people who were interested in food but had no professional stake in food production. Some were motivated by the aesthetics of local food and romantic visions of small farms, while others saw food and farming strictly in environmental terms. Like both of my children, many had become vegetarians at a very early age and were now exploring that interest more deeply. Still others were motivated by the turn toward food issues that was taken by people who started in civil rights, women's issues, or environmental justice. Like me, most of these people had no farm background and had come to their interest in food topics later in life. And increasing numbers of these people were taking a philosophical bent. Philosophers everywhere were starting to use Pollan's books in courses called "food ethics." The time was ripe to write for a philosophy audience for a change.

Initially I envisioned this book less as an intervention into the world of food and agriculture than as a summary of some things that farmers, business people, and food or agricultural might take to be incredibly obvious, but that would *not* be apparent to people who are new to thinking ethically about food. It has not turned out exactly like that. I doubt that very many of my friends in food and agriculture have ever taken the trouble to read what Aquinas said about gluttony in his *Summa Theologica*, for example. In hybridizing some basic philosophy with some basic food and agricultural science I hope that I have come up with something that everyone will find thought provoking and informative.

As usual, there are many people to thank, and no one but myself to blame. A few colleagues read partial drafts or summaries, or reacted to oral presentations of drafts for chapters, sometimes offering extensive critique but sometimes only a brief comment. Both have been very useful. They include Michiel Korthals, Fred Gifford, Stephen Esquith, Sandra Batie, Patricia Norris, Rebecca Grumet, David Schmidtz, Erin McKenna, Darryl Macer, Lawrence Busch, John Stone, Clark Wolf, Elizabeth Graffy, Raymond Anthony, Joy Mench, Ruth Newberry, Janice Swanson, and Kyle Whyte. I am sure that I am forgetting too many. Micaela Fischer, Kaitlin Koch, Zachary Herrnstedt, Nagwan Zahry, Huaike Xu, Zachary Piso, and Ian Werkheiser all read the penultimate draft of the manuscript and offered helpful suggestions. Marion Nestle, Holmes Rolston III, Bryan Norton, and Peter Sandøe offered words of encouragement and a some important last-minute corrections after reading the final draft. I would like to thank Erin Anderson for her work on the figures. Julie Eckinger at MSU has provided constant help with various technical aspects of producing the manuscript, and Lucy Randall at OUP has been a constant source of guidance and support. I must acknowledge my indebtedness to Heather Hambleton, the copy editor from OUP who caught errors and made many suggestions to improve readability, and Molly Morrison, project manager from Newgen Knowledge Works. My appreciation also goes to Ken Marable, who helped in preparing the index. And finally I thank my readers. Giving one's time to someone else's ideas is among the greatest gifts that one human being can bestow upon another.

FROM FIELD TO FORK

Introduction, with a Rough
Guide to Ethics

Dory lives on five acres near a major metropolitan area. She derives most of her income from substitute teaching in several nearby school districts. She likes that work because it allows her to taper off her teaching in the springtime so that she can farm her land. She sells fruits and vegetables to local chefs and to the public at farmers' markets throughout the summer. Dory is especially known for her strawberries, grown without any synthetic pesticides or chemical fertilizers. She has a neighbor who also grows organic strawberries, so occasionally they team up by pooling their berries. One or the other of them will take a turn selling them at the downtown farmers' market. Whether it's Dory or her neighbor Pat behind the market stand, people seek out these berries both for their wonderful flavor and because they like to buy from people they know.

Is Dory doing anything unethical? Many people will be surprised by such a question because the description just given hardly suggests any basis for suspecting unethical behavior. But by selling her neighbor's strawberries Dory is violating the rules for many urban farmers' markets. Although these rules are far from universal, many urban markets limit farmers to selling only the things they grow themselves. Such rules were put in place to give farmers an economic opportunity, but also because people who shop in farmers' markets want to know that they are buying food directly from the person who grew it. Although it might seem unreasonable to apply such a rule so strictly in a case like Dory's, a very similar type of horse-trading among local farmers led to a scandal in 2011 when people in Oregon were sickened by E. coli-contaminated strawberries. The source of the contamination was eventually traced to deer that had been on the farm of a single grower who was supplying numerous roadside stands as well as farmers selling in farmers' markets.[1] The scandal was less the result of the contamination itself than the difficulty authorities had in tracing the contaminated berries through the chain of trades being made by farmers whose customers thought that they were buying direct from the field.

Consider Walker, a student who purchases the meal plan at a college located in an urban area. Walker is a health-conscious vegetarian who eschews most of what is available at lunch and dinner in the campus dining hall, but his meal plan includes a budget for items that can be purchased anytime at the campus snack bar. Walker is not much for snacks, but he can also use this part of his plan to buy a few nonperishable food items: candy, chips, processed meat sticks, and packaged bakery items. One of his similarly health conscious friends has started a campaign for like-minded students to spend their snack budget on these nonperishable goods and then donate them to the local food bank. The director of the food bank says he would love to have them. His clients like candy and chips and especially those peppered sticks of jerky and sausage! But Walker is not so sure. He is all in favor of lending a helping hand to people who are short on food—and after all, he's already paid for the plan, whether he spends the money or not. Yet how can it be ethical to give needy people food that he is not willing to eat himself?

I learned about Walker's quandary by speaking with students at the university where I work, but questions very much like the one he is asking are faced by the managers of local food banks and charitable assistance programs everywhere. Similar questions apply to public policies offering supplemental nutrition assistance programs (SNAP) or what many still call "food stamps." No one thinks that a diet consisting entirely of chips, candy, and soft drinks is healthy. Not even the manufacturers of these foods would suggest *that*. Still, snack foods consumed in moderation can be part of a healthy diet. Denying access to them seems like telling the client of a food bank that they cannot be trusted to make their own choices simply because they are poor or have fallen on hard times. It looks rather like a paternalistic form of disrespect. Yet as the old adage has it, "Beggars can't be choosers." Don't people who contribute to food assistance programs have every right to insist that the churches, government agencies, and charitable organizations who run them shape the program according the donor's values?

And finally, take Camille, a local legislator from a part of the country that is heavily dependent on pork production for employment and tax revenue. Camille has just met one of her constituents, a pig farmer demanding that she support a new piece of legislation. It seems that one of his neighbors hired a college kid to work on his swine farm over the summer, but the kid turned out to be an animal-rights activist. The kid smuggled in some high-priced vodka to drink with a few of the farm's regular employees and then cajoled them into play-acting some scenes inspired by the horrible Abu Ghraib photographs of tortured Iraqi prisoners—only this time with pigs playing the role of the torture victims. The neighbor was furious when he found out. He fired the college kid and docked the pay of his regulars, warning them never to let something like that happen again. But now the video that the kid took of these fake abuse scenes has gone viral on YouTube. The news stations are starting to pick it up and are playing the story as if this is what happens all the time on area pig farms! Camille's constituent wants her to support a law that would make distribution and reproduction of photographs or video recordings obtained without the farmer's permission a crime.

Camille is not so sure. They call these "ag-gag" laws, and versions of them have been passed by state legislatures throughout the Midwestern farm belt of the United States. Although it's easy to sympathize with the plight of her constituent's neighbor (assuming he is telling the truth), such photographs and films are viewed as political speech by the animal protection organizations that circulate them. Camille suspects that her politically conservative farming constituents would not be very sympathetic to government interference in their own speech. And how can she (or anyone) be sure this neighbor *was* telling the truth when he accused the college kid of filming a set-up? Maybe those pigs were actually being abused. At the same time, she's troubled because—having campaigned on plenty of pig farms herself—she's satisfied that even if this kind of abuse happens from time to time it is rare. But the whole industry suffers when pictures like this are made public, and where is the justice in that?

Dory, Walker, and Camille are struggling with tough questions in food ethics, but many issues in food ethics are not tough at all. In 2009, Chinese officials revealed a conspiracy in which infant formula had been deliberately adulterated through the substitution of melamine, an ingredient in industrial glues and plastics, for milk powder. At least three infants died as a result, and some estimates indicated that 300,000 were sickened and may experience long-term health consequences. The perpetrators of the conspiracy are believed to have gained millions of dollars, but at an intolerable cost in human misery and loss of life.[2] Unlike the questions being posed for Dory, Walker, or Camille, there is no mystery here about what should be done. Yet as much as we might like to think that the matter of what we eat or how it is produced and distributed will always be simple and clear-cut, the preparation and consumption of foods we eat everyday are replete with opportunities for ambiguity, confusion, and disagreement. Some of the most enduring and deep disagreements occur when one person thinks the ethical choices are easy and unambiguous, but the next person is not so sure.

The Rise of Food Ethics

Dory, Walker, and Camille are philosophical thought experiments—stories cooked up to give us insight into an ethical problem—rather than real people. Their situations are typical of problems that will be discussed throughout this book. To many people, food ethics means making better dietary choices. Choices could be better in terms of health or they could have better environmental and social consequences for others. Food choices become ethical when they intersect with complex economic supply chains in ways that cause better or worse outcomes for other people, for nonhuman animals, or for the environment. It is worth reminding ourselves that this is a relatively new idea.)Enthusiasm for farmers' markets; humanely produced animal products; and fairly traded coffee, tea, and cocoa has grown markedly over the last decade. Over the same time period, we have also gained greater recognition of links between diet and the alarming growth in diabetes, heart disease, and other degenerative conditions. Thus food ethics might include not only making better choices yourself but also designing menus, public policies, or even cities to encourage better food choices by everyone. The examples of Dory, Walker, and Camille illustrate further problems in food ethics that do not even involve dietary choice in any simple or straightforward way.

The growing number of ways that food becomes embroiled in ethical quandaries coincides with key industrial and commercial developments in the production and distribution of food. As food historians have demonstrated, the early decades of the twentieth century saw the emergence of food manufacturing firms and chain grocery stores. During this period, many factors conspired to create a food system in which consumers were quite ignorant of where their food came from and hence *could not* make choices on ethical grounds. On the one hand, urban populations simply lacked a kind of personal experience with food production that had been virtually ubiquitous a century earlier. On the other hand, technological changes in rail transport and food processing were creating longer supply chains and smoothing out seasonal variation in food availability. Branded products arose in response to consumer demands for some reasonable certainty as to the quality of processed foods, and with branding came food advertising. Home economists promoted the use of canned and packaged food as "progressive," and as more women entered the workforce, it became necessary to economize on the time invested in procurement and preparation of meals at home. By the 1960s, these trends were being augmented by rapid growth in meals eaten outside the home.[3]

These developments in marketing and distribution were occurring as a two-centuries-long process of transformation was being completed in the agricultural sector of industrialized economies. The years after World War II were especially significant for a rapid growth in the use of chemical methods for controlling plant diseases and insect pests in crop production, and the creation of intensive concentrated animal-feeding operations (CAFOs) or, as their critics characterized them, "factory farms."[4] The combination of consumer ignorance and complex technological change began to be associated with a series of high-profile problems, beginning with food adulteration during the early decades of the twentieth century and continuing with Rachel Carson's exposé *Silent Spring* in 1962. The consumer backlash began to mount in the counterculture with increased attention to the health and environmental impacts of industrially produced and distributed food. Sometimes this backlash took the form of small farms and food co-ops that attempted to create an alternative food system, but the more typical response has been for economically successful farmers and well-entrenched food industry firms to develop and market products that appeal to "alternative" values. In the 1970s, foods were advertised as "natural," but consumers rapidly turned to "organic" as a more meaningful alternative value. Soon "humane" and "fairly traded" labels began to be added to the alternative food lexicon. By the early years of the twenty-first century, consumers who had been made skeptical by large food industry firms' abilities to exploit all these terms began to seek "local" foods as a way to eat more ethically.[5]

This book engages these topics, but it does not tell you what to eat. Chapter 1 reviews the recent history for the rise of food ethics in more detail, emphasizing seminal ideas that emerged in the 1970s. One theme is to describe the difference between food ethics focused on supply chains and their socio-environmental impact, on the one hand, and an ethics constructed wholly in terms of one's own dietary choices, on the other. From there, the book takes a deeper dive into a few of the big themes in food ethics. Injustice in the food system is the focus of Chapter 2. I ask whether food issues tell us something new about social justice, or if they are simply case studies for more general philosophical ideas about justice. Chapter 3 follows the ethics of diet and health from the ancient world to our current obesity crisis. In Chapter 4 we come to what I call the "fundamental problem" in food ethics: the enduring tension between the interests of poor farmers in the developing world and the hungry masses of growing urban centers. The case for vegetarianism is discussed briefly in Chapter 5, but the main focus is on the ethical difficulties in food animal production that we should be thinking about while we wait for people to become vegetarians. The environmental sustainability of the food system is the subject of Chapter 6, and Chapter 7 takes us back to the developing world to consider how Green Revolution-style development projects should be evaluated in ethical terms. The final chapter revisits themes of risk, personal diet, and the nature of ethical thinking itself by considering some questions in the debate over genetically engineered foods. These eight chapters provide a microcosm of the issues that *might* be included under the heading of "food ethics." In each case I do more to complicate the ethical analysis of our contemporary food system than I support specific recommendations for policy change or personal choice. There are many more topics that could be added.

What Is Ethics?

I believe that ethics should be viewed as a discipline for asking better questions. Common speech equates "ethics" with "acting rightly," and like many college professors who teach and write on ethics, I am frequently beset by someone who recounts an episode of outrageous behavior by some person or group. The tirade ends with the question, "So is that ethical?" It seems as if they want an expert to certify their opinion. When the behavior in question is truly outrageous, it is easy enough to agree, but philosophical ethics is more attuned toward developing the vocabulary and patterns of thinking that make for more perceptive and imaginative ethical reasoning than it is toward training someone to judge particular cases in a uniform manner. In this section and the next, I provide a brief introduction to the way that philosophers approach ethics. It is intended to ease readers who come to the book with a keen interest in food issues but little background in philosophical ethics. These remarks provide a sketch and cover some standard terminology for lay readers, but take caution: in answering the question, "What is ethics?," this way, I do not pretend to be offering definitive accounts of the various schools or theories of ethics that are studied in philosophy departments.

We can start with a graphic that illustrates some key elements of a decision-making situation in very simple terms. Figure 1 represents the *agent*, a person or group that will undertake some action or engage in some kind of activity. One can think of oneself occupying this position, but the figure might also represent an organization, such as a business or political party, or it might even be taken to represent society as a whole. The activity to be undertaken may or may not be the result of a conscious or deliberately considered and calculated choice. The situation of a very generalized "decision maker" orients us to some key elements of human conduct that can be the target of ethical reflection or evaluation. After noticing these elements we reflect equally on the actions that we undertake as individuals or on the combined activity of people acting in organizations, groups, and even random groupings (such as crowds).

Human activity is always constrained in a number of different ways. Some things are simply impossible: they violate the laws of physics and chemistry. One cannot become invisible on whim, and in the world of food, biophysical constraints define some of the possibilities for the production and distribution of what people will eat. In the graphic, such constraints are represented by the ring labeled *technology*. In referring to these biophysical constraints as technology, we acknowledge that although the laws of physics, chemistry, and biology limit what can be done rather robustly, the way that these inflexible laws are reflected in our material circumstances is *not* fixed once and for all. The history of food and agricultural technology has dramatically changed the biophysical constraints that

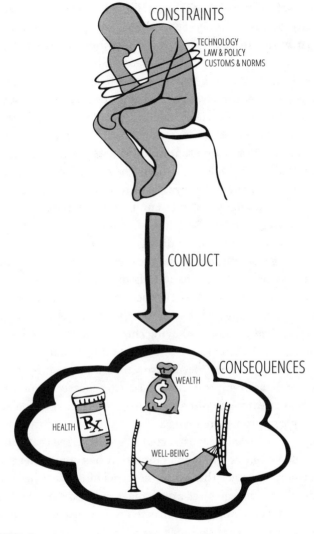

FIGURE 1

someone living in the nineteenth century would have faced. Some key issues in food ethics concern the way that we should invest research dollars in the quest to make even further changes in our food technology. However, the main focus of ethics is usually on the two "softer" types of constraint. First, there are things that are forbidden by *law and policy*. Some things are against the law, while others may violate a policy that one is required to follow as a condition of employment, for example. Second, there are biophysically possible courses of action that are forbidden by *customs and norms*. While not against the law, they are things that one knows full well not to do. In typical decision making, people have so thoroughly internalized both kinds of constraint that they simply do not even consider a course of action that would violate them.

Eventually, people do something; the agents act. Perhaps they rather unreflectively order lunch at a restaurant, or perhaps they make a life-changing decision to quit their job and start an organic winery in Oregon. A company acts when it launches a new product or advertising program. Even disorganized groups can act. A riot would be one example, but we say that "the public" acts (or speaks) when a new product succeeds or fails to attract enough buyers to achieve market success. Whatever action or activity is performed or undertaken is referred to as *conduct* in the graphic. It might be cooking mashed potatoes or buying chips at the store, and it might be passing new laws on food safety or making a multi-million dollar investment in some speculative venture to produce test-tube meats. For firms, it might be manufacturing cheesecake or marketing a product through online sales. Any description of something that people or groups are doing could qualify as conduct, however broadly or narrowly it is constructed. We start to see that this very general depiction of the human situation has some relevance to ethics when we notice that all these forms of conduct have *consequences*. Here, consequences are changes or effects on the health, wealth, and well-being of any affected party, including the agent herself (or itself, as the case may be). From the perspective of clear ethical thinking, it is important to distinguish the conduct from its impacts or consequences. The sum total of *all* consequences to the health, wealth, and well-being of all affected parties is often referred to as an *outcome*. It would be possible to develop more detailed or nuanced terminology for describing ethically significant action or activity, but this very simple model will suffice for present purposes.

Each element of an agent's decision-making situation can have ethical significance, and the nature of this significance is often signaled by distinctive terminology. Ethically significant consequences are described as benefits, costs, or harms. If an agent does something that affects the health, wealth, or well-being of someone else positively, it can be characterized as a benefit. If it affects that person adversely, it is a harm. We commonly refer to adverse impacts as *costs*, but this is not how economists understand that word, so there are reasons to avoid it. Ethically significant constraints that take the form of law and policy, on the one hand, or customs and norms, on the other, are often described in terms of *rights* and *duties*. If someone has a right that I must respect, doing so constrains my potential range of action. Possible courses of action that would violate that person's rights are considered out of bounds and I must not undertake them. If I have a duty (perhaps because I have made a contract or promise) my action is similarly constrained: I can consider *only* those possible activities that are consistent with fulfilling that duty. Generally speaking, rights and duties can be thought of as correlative: if I have a right, then others have a duty to respect it. Finally, there are certain types of conduct that are named directly with words that imply ethical significance. *Lying, mendacity,* and *dishonesty* are rough synonyms for one type of conduct; *truthfulness, honesty,* and *sincerity* name its opposite. Such words tie ethical significance directly to a given type of conduct without referring back to rights and duties *or* looking ahead to outcomes. They classify conduct in terms of virtue or vice.

Philosophical ethics is an organized practice—a discipline. Its practitioners focus first and foremost on these three ways that action or activity can be characterized as ethically significant. They study the way that people formulate, specify, and discuss the rightness or wrongness of action and activity by characterizing it in terms of virtue or vice, noting the constraints that function as rights or duties, and attending carefully to its beneficial and harmful consequences. It is only a slight exaggeration to say that if a point of reference *cannot* be described in terms of relevance to virtues and vices, rights and duties, or benefits and harms, it is unlikely to have any ethical significance at all. Food ethics, then, is the study of how virtue, vice, rights, duties, benefits, and harms arise in connection with the way that we produce, process, distribute, and consume our food.

How Philosophers Approach Ethics

For the last two hundred years or so, academically trained philosophers have tended to organize themselves among several schools of thought on how actions and activity can be deemed ethically correct or, to say the same thing, ethically justified. They have developed theoretical accounts of what makes one action right and another wrong, and these accounts are referred to as *ethical theory*. Some of the most influential theories focus on only *one* of the elements described above. Utilitarianism is a theory (actually a family of theories) holding that ultimately only consequences matter. Claims about rights and duties are, at the end of the day, reducible to benefits and harms. A duty, for example, might simply be a rule that, if followed carefully, will lead to the best consequences—the best available outcome in terms of total benefit and harm. Classical utilitarianism specified an ethically right action as the one that achieves the "the greatest good" (i.e., the most net benefit) for "the greatest number" (i.e., the largest possible number of affected parties). This specification was modified by welfare economists who were willing to sanction any action that achieves a "potential Pareto improvement" (i.e., achieves more benefits that harm). Notice also that this simple description of utilitarianism does not say *who* or *what* counts as an affected party. Is it only one's fellow citizens or is it all of humankind? What about nonhuman animals? Is it possible to benefit or harm an ecosystem or an endangered species? There are other details, too. Clearly, not *every* adverse event counts as a morally significant harm. If you beat me at Monopoly or Scrabble, it's doubtful that I can claim to be harmed in any morally significant way. But what about the losers in an economic competition? Is losing your job because your employer went broke just an example of "the way things go" (like losing at Scrabble), or is it a morally significant form of harm? Here there are disagreements, and as we will see at several junctures, this disagreement matters in food ethics. For now, just note that spelling out the details of a full-fledged ethical theory becomes an exceedingly complex task.

For present purposes it is the contrast between a utilitarian's laser-beam attentiveness to consequences and an approach focused on rights and duties that is more significant to notice. Such a theory, which we will simply call *rights theory*, must derive an account of the way that actions are constrained by the rights of others. Although (once again) this can get very complicated, it is worth pointing out two general strategies. One approach is called *contractualism* (or sometimes *social contract theory*). This approach assumes that rights are grounded in the promises that we make to one another. If I promise to meet you at the pub at five o'clock, you have a right to expect me to keep that promise, and I have a duty to do so. When we (or our representatives) make laws or set policies, we are, in effect, promising to act according to a system of rights and duties. More generally, perhaps we can think of our social interaction as a set of implicit promises, and we can develop our ethical theory by asking each other, "What are the promises that we would most hope to govern our interactions?" Historically, this question has often been tied to rationality: what is the social contract—the system of rights and duties—that a rational person would accept? The contractualist approach emphasizes the *reasons* why a given social bargain (i.e., a system or configuration of rights and duties) would be rationally acceptable, rather than the benefits and harms that such a configuration can be expected to produce.

The alternative approach to rights suggests that we can derive a binding set of rights and duties by thinking hard and deeply about the nature of human freedom. The widespread practice of human slavery was overturned largely because people came to believe that it could not be reconciled with basic human rights. Of course a truly free person must not be a slave to passion, either. And passion is governed (or constrained) by a good or moral will. On this view we can obtain a sense of mastery over passions that destroy our freedom—a perspective on freedom indicated by the term *autonomy*—by recognizing (or in some sense giving ourselves) a set of constraints to guide our conduct. Immanuel Kant developed the most influential version of this approach to ethical theory, arguing that we can impose correct moral constraints upon ourselves by asking whether we would be willing to see a proposed constraint treated as a universal law—as a principle of duty binding on all persons at all times. An approach to rights theory that probes the meaning and achievement of human freedom is often described as *Kantian* (or *neo-Kantian* in deference to some deviations from Kant's own view). Again, fully specifying such a theory brings us into further complexities. The point here is definitely not to convey an adequate basis for understanding neo-Kantian ethical theory but simply to indicate how someone pursuing this line of reasoning will ask rather different questions than someone who thinks that ethics can be satisfactorily theorized by calculating the net value of benefits and harms.

And there are still more options. The type of philosophy that gives rise to debates over rights and duties on the one hand and consequences on the other fails to really capture all of the meaning that is sometimes packed into the claim that a particular type of conduct is virtuous or vicious. While talk of the virtues seems less able to drive ethical thinking to a specific prescription—that is, to a formulation of which action really is the right thing to do—it nonetheless *does* articulate the way that patterns or habits of conduct become morally significant, even when they are undertaken relatively unreflectively. Emphasizing virtue (or *arête*, as Aristotle might have had it) may be especially useful when an overall pattern of behavior rather than a single instance of choice is morally significant. Some advocates of *virtue theory* emphasize an individual's disposition or character, not whether any given decision conforms to a rule. Alternatively, achieving virtue may depend upon living in an environment or culture that shapes our behavior in ways that we are barely aware of. Promoting virtue may have more to do with structuring human behavior and social interaction in ways that make it easier to be reflective about the things that really matter and that steer us onto autopilot when proper action can be reliably left to habit. So some philosophers have defended virtue theory as an alternative to relying on close inspection of consequences or rights that arise in connection with any single instance of action.

There is less than total agreement among contemporary philosophers that choosing one of these tracks in ethical theory is required at all. One might be a pluralist who sees each way of thinking as achieving a kind of partial truth. Or one might follow Jürgen Habermas, who argues that we should be focused on the *process* of engaging these different types of ethical reasoning in a form of discourse or debate where discussants trade arguments in the spirit of reaching a kind of agreement on what is right for the case at hand (a view Habermas calls *discourse ethics*). Perhaps I should admit to being more fully persuaded by Habermas than by any contemporary advocates of utilitarianism, rights theory, or virtue theory. Suffice it to say that I will *not* pursue detailed development and application of *any* of these theoretical approaches in the following excursion through food ethics. Nevertheless, it will be helpful to readers to be attentive to the way that ethical arguments can function in these three somewhat different ways. Sometimes a reason for doing one thing rather than another is based on the outcome of action, and sometimes it is grounded in the way that action should be constrained, either by social convention or by the nature of our desire for true freedom. Sometimes a reason for doing something or even for thinking harder about what one is doing out of habit appeals to a more nebulous but nevertheless palpable sense of virtue. All these types of reasoning make occasional appearances in food ethics, and the point of this brief summary is simply a heads-up to the reader unschooled in the ways of the philosophers.

A Note on My Method

My approach in this book aims to steer a path between the Scylla of always keeping your peas and mashed potatoes separate and the Charybdis of mushy thinking rationalized by whatever seems right at the moment. In other words, I deploy the philosopher's penchant for clear and distinct ideas, but I deploy it in moderation. I treat ethical theories (such as rights theory, utilitarianism, or natural law) as argument forms that provide alternative (but sometimes also complementary) ways to frame a descriptive account of the situation that confronts us. In doing so, these accounts make claims upon the emotions, habits, and institutional structures that allow us to act as individuals, as informally coordinated groups, and as formally structured organizations. I take it that one job for ethics is to inspect and query the circumstances in which such claims arise. Although these claims are often made explicitly by people who wish to motivate action, I do not presuppose that the claims upon us have always been clear or explicitly articulated.

I do suppose that in investigating ethical issues in the food system over the last thirty-five years, I have been engaged in inquiry. By *inquiry* I mean a loosely structured activity that arises in reaction to some disturbance or disruption and that expects to conclude with an active response that resolves or at least responds to the distress or curiosity with which the inquiry began. I *have* been trying to think of these issues in the right way. However, in pledging my allegiance to getting it right I am not *also* promising to frame matters in terms of ethical *or* scientific theories. I am not intrinsically interested in portraying the issues as social constructions or functions of underlying biological drives, to note just two of the many ways that social or biophysical scientists treat food issues. There may be occasions in which either of these modalities are helpful to my inquiry, but I am not here to peddle theoretical constructs. In undertaking ethical inquiries, I am hoping to arrive at better and more correct answers than I started with. One influential tradition in philosophy has supposed that I must have some prior conception of what it means to get things right in order to do this, but my own view is that any conception of what it means to get things right would *itself* be the product of an inquiry. So I am inclined to think that in concluding this introduction it may be more important to say something about inquiry itself.

John Dewey illustrated his basic conception of inquiry in the 1896 article "The Reflex Arc Concept in Psychology." Sitting

THE LEARNING CYCLE AND THE FOUR PHASES OF INQUIRY

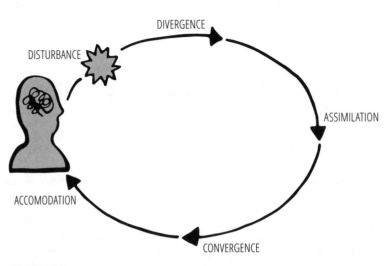

FIGURE 2

comfortably and engrossed in a book, he is startled by a noise. His first response is a bit scatterbrained as he reorients his attention to the disturbance that is making a claim on his attention. He forms a hypothesis: the wind has blown the window open. Next he forms a plan: get up and close the window. Finally, he actually gets up and closes the window, bringing the inquiry to a close through an action that simultaneously corroborates the hypothesis, executes the plan, and addresses the disturbance. For Dewey, it is the whole sequence that illustrates inquiry, *including* the scatterbrained part and the physical activity of closing the window itself.[6] The educational psychologist David Kolb offers some helpful terminology and a schematic of Dewey's learning theory that identifies four distinct phases—five phases if we count the disturbance that sets the whole thing off (see figure 2). The scatterbrained search for a general orientation is *divergence*. As the dizzy and unfocused divergent stage begins to coalesce into a more structured and organized search for answers, the hypothesis-forming or *assimilation* phase begins to construct an explanation or model that accounts for the disturbance. Once this is in place, it is possible to formulate conditional if-then components that, in turn, suggest a plan, a process that Kolb calls *convergence*. Finally, it is worth noticing how executing the plan requires many elements that were probably not anticipated by the hypothesis. One has to actually get up out of the chair, which may require finding a place to put one's book. Closing the window may require jiggling or bumping it. These supplements to the hypothesis involve an active and engaged kind of intelligence that Kolb calls *accommodation*. A learning cycle is completed when a person or group has moved through all phases of the inquiry process.[7]

A detailed discussion of Dewey or Kolb would take us far from food ethics, yet it may prove helpful to notice how this four-phased account of inquiry helps us orient ourselves to a number of tasks that are crucial for practical ethics. Each phase in the process of inquiry is associated with a distinct potential for error, for mistakes that will result in a failure of the overall process. In the divergent phase, it is important to keep the opportunities for brainstorming and bringing possible responses to the disturbance open in order to avoid over-commitment. An unfruitful investment of cognitive resources (e.g., time and energy) into a hypothesis that turns out to be unrelated to the disturbance or curiosity that the process of inquiry was initiated to address is one kind of error. It is, in fact, an all too frequent kind of error in today's world. In assimilation, the goal of explanation or modeling takes over. Assimilation points us toward classic philosophical characterizations of truth and moral correctness. To be in error here is simply to have a model or explanation that does not accurately map or correspond to the situation. Twentieth-century moral philosophers became obsessed with modeling a universal standard for moral correctness, so much so that for many philosophers, "getting it right" came to be understood exclusively in terms of limiting assimilative errors. As one moves into convergence, a number of more practical considerations begin to be relevant. There may be many ways that one *could* undertake action given the working model that has been developed in the assimilative phase, but it would be a mistake to ignore the relative costs and risks that would be associated with any given possibility as one converges toward implementation. In the accommodation phase, it becomes important for somebody to actually get up and close the window. It is here that the classic divide between theory and practice emerges, for the proverbial "man of action"—quite possibly a woman, I hasten to add—may be the person or group who is most able to avoid the distracting tendency to revise the theory instead of finally doing something. The accommodator adjusts the plan so that the initial disturbance is materially addressed in its particulars.

This discussion of errors suggests that a strictly sequential interpretation of these four phases of inquiry may be misleading. Indeed, Kolb himself has stressed the idea that different individuals may be characterized by a particular learning style. Each learning style is typified by different types of intelligence, capability, or skill. The accommodative learning style encompasses a set of competencies that Kolb calls *acting* skills: leadership, initiative, and action. The diverging learning style is associated with *valuing* skills: relationship, helping others, and sense-making. The assimilating learning style is related to *thinking* skills: information gathering, information analysis, and theory building. Finally, the converging learning style is associated with *decision* skills like quantitative analysis, use of technology, and goal setting.[8] Kolb argues that in the university we see these different learning styles strongly associated with different academic departments. Academics with diverging learning styles tend to become English professors or teach in the arts. The natural sciences are dominated by assimilators, while engineers and others in technology fields tend toward a convergent learning style. Accommodators wind up in the business school.[9] Without endorsing the stereotyping implied by such classification of individuals, it is nonetheless striking that a lot of work by academic philosophers tends to valorize criteria that reflect the assimilation phase of inquiry—an approach that Richard Rorty critiqued as seeking to become "the mirror of nature." My approach in this book is pragmatic in holding (with Dewey) that it is the totality that should remain foremost in our thinking while undertaking a process of inquiry. There are many ways in which we can err, and there are numerous ways in which we can get it right.

Kolb's schematic is most relevant to our topics in that it helps us recall that criteria for right thinking and right action will depend on where we are in the process of inquiry. It suggests a picture of learning or inquiry that errs when any one of these learning styles comes to dominate our thinking, and in this it intersects nicely with recent trends in feminist and postcolonial epistemology. We are not likely to get things right when we systematically exclude people who have a particular perspective from the processes of deliberation and social decision making. There is a growing recognition among people from many walks of life that this kind of exclusion is not only unjust, it is spectacularly stupid in its tendency to discard or ignore what may turn out to be crucial pieces of information. This has resulted in a wave of philosophical and social science research that explores the process of inclusion. Here, too, there is an important intersection point with the approach I take in this book. Social action to address the issues discussed in the following chapters must clearly take cognizance of this work and must experiment with more inclusive modes for organizing and affecting responses.

However, right action in food ethics will likely require a bit more than a general theory of inclusion or social process. Consistent with feminist epistemology and critical theory, my approach to food ethics does indeed emphasize inclusion and listening. Consistent with an interest in social justice and with at least some goals that have bound people into a food movement, the following chapters probe a series of food issues with an eye toward identifying the key points at which divergent social concerns meet. Consistent with the pragmatist orientation, the analysis seeks to identify the points at which divergent, assimilative, convergent, and accommodative learning styles intersect. Consistent with yet another line of recent scholarship in science studies, my discussion of food and food systems will focus on the objects, organizations, and activities that reside along the boundaries of these intersections. But consistent with all of the above values, I largely ignore the theoretical apparatus in favor of plain talk, whenever I can do so. I hope that I uncover unnoticed and underappreciated sites where action might be focused, and that I identify some divergent themes that need to be recognized before we prematurely converge upon plans of action. Ironically perhaps, it is not necessary to call a great deal of attention to the theoretical and methodological themes of epistemology, pragmatism, and moral theory while doing so.

Chapter 1

You Are NOT What You Eat

The way you cut your meat reflects the way you live.

Confucious

Tell me what you eat and I will tell you who you are.

Jean Anthelme Brillat-Savarin

Man is what he eats (Der Mensch ist, was er ißt.).

Ludwig Feuerbach

You are what you eat.

United Meat Market, Bridgeport, CT

You are what you eat. The aphorism enjoys recurrent popularity. It reinforces myriad types of dietary advice: "Eat your vegetables." "Eat your meat." "Drink your milk." "Chew your food." "Don't slurp your soup." "Clean your plate." "Don't eat that junk food." "Don't be so picky." "Don't eat anything that can look you in the eye." "Eat more." "Eat less." Each message is underlined by adding, "You are what you eat." In ordinary language, the phrase works almost like an exclamation point, but it also conveys ethical force. In each case, it is understood that one is saying "Eat what you should," but notice that *what* one should eat varies from one context to another. An Internet database indicates that the phrase (or some clever permutation) has appeared in the title of scientific or scholarly articles several dozen times in the last decade. For Confucious and Brillat-Savarin, the point seems to have something to do with social class or lifestyle. For Feuerbach, it prioritizes biophysical existence over any putative spiritual being. The philosophers are making ontological claims: they are saying something about what kind of thing a human individual is. But in contemporary usage, the expression is almost always a dietary recommendation.[1]

Does the assertion that one is what one eats become less philosophical when detached from the ontological concerns of Confucius or Feuerbach? Is there any sense in which dietetics—a set of norms and guidelines governing food consumption—can be thought of as an *ethical* discourse? The twentieth century was notable for its relative absence of philosophical reflection on food, though Michel Foucault remarked on the appropriateness of such reflection. But a number of studies in the 1990s suggested that philosophers were poised for a rebirth of interest in food studies and now at the midpoint of the 2010s, it looks as if they were right.[2] This book examines just a few dimensions of this rebirth, all centering around the theme of food ethics. My review includes both philosophical and extra-philosophical writings.

The reawakening of food ethics can be attributed to cultural, political, and social changes, as well as to the late twentieth-century development of some key philosophical doctrines. One of remarkable things about food ethics is its capacity to cross boundaries and span a surprisingly complex array of topics. I make no pretense of exhaustive coverage, but this chapter begins with a brief survey that is intended to illustrate the breadth. There are issues of social justice, then impacts on animals and the environment. Eventually questions about risk to health as well as risk to cultural traditions arise as well. Before moving on to subsequent chapters that examine these themes in more depth, we conclude the overview by offering a preliminary answer to the question of whether food practices can (or should) be regarded as having ethical significance. The answer is that it all depends on what you mean by ethics. There are cases such as the Chinese food safety debacle in which food ethics concerns the way that one person's or group's conduct affects others. But in terms of the way that philosophers have construed the subject matter of ethics for about two hundred years, the performance of cultural foodways is different. Given this tradition, acts of food production or consumption that support bonds of sociality or create personal identity are ethically significant in much the same manner as aesthetic or religious practices. They are important for insiders, but are not to be imposed on others. That may be changing, however, and that is precisely why food ethics could have broader philosophical significance for the way that we understand ethics itself.

The Long Eclipse of Food Ethics

Although I speak of a *return* to food ethics, very few of the philosophical texts we now regard as emblematic of the last five centuries discuss food at any significant length. Mention of food and eating is most likely to occur in examples or metaphors oriented to other topics. For example, in order to illustrate a point about aesthetic taste, David Hume refers to the taste that a leather-strapped key left in the bottom of a cask conveys to a glass of wine. As evidenced by the three quotations in this chapter's epigraph, there were a few attempts to reference eating while reflecting on the ambiguous social and ontological status of human beings. The observation that people from different social strata eat different things hardly seems profound. Somewhat more thoughtfully, humans seemingly inhabit an immaterial domain in their capacity as intentional actors or thinkers, yet they are also embodied beings who exist only insofar as they partake of material sustenance. Pondering this dualism has been a source of fun for philosophers ever since Descartes. Feuerbach and Nietzsche were challenging Descartes' dualism and the Christian view that mind is more important than matter when they said "You are what you eat." But this kind of parrying is still far from a genuine philosophical concern with dietetics.

A number of philosophers since 1500 have taken some interest in the production of food. Before that time, soil and climate were thought to provide weighty impress on the formation of culture, personality, and political institutions. People from tropical climates were thought to be "hot tempered," for example, or the absence of fertile soil was thought to reinforce an inconstant and nomadic culture. This way of thinking anticipates a Darwinian or evolutionary view in one sense: the natural philosophers who advanced theories on the determinative powers of soil and climate took a keen interest in the way that organisms (especially humans) were adapted to a biophysical environment.[3] Eventually, philosophizing about the powers of soil and climate gave way to debates over the political economy of food production. Here philosophers probed the way that incentives and constraints on the material reproduction of the populace were tied to an ability to feed them (or at least their armies). In political economy we have a pattern of inquiry running from Machiavelli and Montesquieu to Malthus and Mill in which the means of food production are examined first in relationship to institutions of state finance and governance, and eventually with respect to the fairness of distribution across social classes. We also have Marx referring elliptically to eating in ponderous passages from the *Grundrisse* where he discusses how for the individual worker, "production is consumption; consumption is production." He meant several things at once, but one was that the exertion of physical labor in work (i.e., production) consumes the laborer by depleting his or her physical strength. The body of the worker is then restored (or produced again) by eating. Although all these philosophers are taking food to be more than a metaphor, they have little interest in *what* gets eaten and they understand food's significance narrowly in terms of meeting the barest requirements of subsistence.

All the same, it is not as if diet and philosophy are simply independent domains. There *is* a tradition of philosophical dietetics that extends back to Pythagorean injunctions against eating beans. Ancient philosophical texts often included advice about what, when, how, and how much to eat. Such advice was deemed a natural element of ethical treatises dedicated to what Foucault has called "the care of the self." In these works, diet, sexuality, friendships, meditation, and physical exercise were included as elements of the routine or discipline that constituted living well. In the Greek and Roman philosophical lexicon, living well and being *seen* to live well were not sharply distinguished. Even the most trivial aspects of a person's daily routine could be regarded as evidence of that person's wisdom or foolishness. Profligacy in any domain would be a sign not merely of laxity but of error and ignorance. For as Socrates taught, how could someone who truly knows the truth fail to act on it? As wisdom was not, for the ancients, Balkanized into spheres of specialized knowledge, the failure to eat well could be taken to be a moral failure of the most general kind.[4]

The Greek virtue of temperance became the framework for philosophical dietetics in the medieval world. St. Thomas Aquinas formulated the sin of gluttony in terms of five more specific vices: *nimis*—eating too much, *praepropere*—eating too soon, *ardenter*—eating too eagerly, *studiose*—eating too delicately, and *laute*—eating too sumptuously. The implied recommendation of a simple diet was swept up in later theological debates that split the Christian world over evangelical poverty. Michael of Cesena, a Franciscan friar who argued that a true imitation of Christ required the abnegation of property, also defended the view that food was abused when eaten for gratification of personal preferences or in excess. Such views put Michael on a collision course with a papacy engaged in increasingly conspicuous consumption. In this respect, Michael's conflict with Pope John XXII anticipated the Protestant Revolution initiated by Martin Luther two centuries later. While it would be exceedingly misleading to suggest that food ethics was a central issue in these struggles, theological dietetics was nevertheless viewed as a philosophically legitimate component of them.

Such themes are not wholly absent from philosophy in the modern period. Johann Gottfried Herder seemingly stands the medieval approach to eating on its head. He advocated a diet of fine—as opposed to course—food as a precondition for ascent into rationality and civilization. Kelly Oliver has explored his preoccupation with diet as an element of Herder's presumption that being fully human involves transcending (or overcoming) animal being.[5] Others in the modern period focused on dietetics in a way that resonates more closely with contemporary debates, arguing that consumption of animal flesh is both unhealthy and immoral. Tristram Stuart puts forward evidence that Descartes, Bacon, and Newton each experimented with vegetarian practice and may well have held philosophical views supporting abstinence from meat consumption. Yet Stuart's history of vegetarianism from 1600 onward also highlights the way that matters of diet generally *failed* to be taken up as philosophical topics. If Descartes, Bacon, and Newton held philosophically grounded views on vegetarianism, they edited such views out of their philosophical writings.[6]

Why then did dietetics vanish from philosophy for five hundred years or more, and why is it coming back now? Clearly the causes for the initial disappearance are complex. As Carolyn Korsmeyer argues in her book *Making Sense of Taste*, Western philosophers have had a prejudice against the bodily senses of touch, smell, and taste that can be traced all the way back to Plato and Aristotle. Disputations on rationality have always relied on metaphors of seeing, with occasional deference to listening. Knowledge obtained through touch, taste, or smell was never put forward as an exemplar of rationality. Although dietetics existed as a philosophical topic in the ancient world, exploring the feel, smell, and taste of food has never been a high-status philosophical avocation. Korsmeyer suggests that as philosophy in the modern period focuses more and more intently on rationality, it is hardly surprising that concern with bodily functions fades ever more deeply into the background.[7]

In addition, there was also a tendency for philosophy to shed topic areas from its portfolio as they became the object of organized scientific research. Astronomy and physics ceased to be the subject of philosophical treatises as they become independent disciplines in the natural sciences. Much the same thing has happened to dietetics. As nutrition was becoming a topic for science, food choices came to be seen as matters of prudence and personal taste rather than morality. Although actions that affect only our own interests can be spectacularly stupid and ill conceived, they cannot be immoral or unethical unless one can *also* find something that connects the damage to our own interests with harmful impact on (or harmful intentions toward) affected parties. In saying this, one does not rule out the possibility of seeing the care of oneself in religious or existential terms. It has not been unusual for Christians to

liken the human body to a temple, for example, then to conceptualize dietary and other health-related practices as a duty to God. But
as contemporary ethics has come to understand such self-imposed
obligations, one accepts such duties because one chooses to affiliate
with a given religious tradition, or because one chooses to see one's
life as a project of a certain kind.

John Stuart Mill's essay *On Liberty* may provide one of the
most persuasive arguments for thinking that it is the effect of
one's action on other people that supplies the moral force for constraints on personal choice. Mill begins with an extended defense
of the social benefits of "freedom of thought," which include law
and policy that protects the free expression and discussion of ideas.
Save for the obvious cases like shouting "Fire!" in a crowded theater,
speech doesn't hurt anyone, while attempts to restrict speech retard
innovations and beneficial social change. And stifling the exchange
of ideas generally serves to protect entrenched groups who are ben-
efiting at the expense of others. Mill then generalizes the argument, suggesting that imposing restrictions on any form of conduct
is fully justified when the conduct causes harm to others. But in
other instances, we should mind our own business. There are some
tricky cases where there may be a basis for intervention in order
to prevent someone from harming themselves (e.g., Mill discusses
drunkenness), but for the most part we should aim for a morality
that restricts a person's or group's liberty to act as they deem fit only
in those cases where the action will inflict harm on another party.[8]

Contemporary philosophical ethics has been committed to the
view that individuals should have freedom in developing their own
life plans. Ethical issues arise when these plans clash, but not insofar as individuals acting from different plans can live compatibly.
In practice, the distinction between prudential or religious duties,
on the one hand, and morality, on the other, has never been so tidy.
Yet *if* one accepts the view that self-regarding interests raise no
moral but only prudential or existential concerns at face value, the
matter of what one eats will tend to be seen as personal. Given this
orientation, food ethics might include the way that food production or distribution affects the welfare of others, but it would not

be concerned with dietetics, with choices about what to eat. Why, then, would a new moralized dietetics arise?

One possible explanation is a widespread cultural devolution to the religiously charged philosophical frameworks of the ancient and medieval world. That is to say that we might be returning to ways of thinking in which prudence and morality are *not* distinguished, and where virtually any act can be seen as a test of moral character. In 2011, portly New Jersey governor Chris Christie withdrew from competition for the Republican Party's nomination for president of the United States after commentators questioned whether he was too fat to be president. While in some instances the comments called attention to prejudicial attitudes in the American electorate that limited his electability, others were quite willing to make the moral claim that someone who cannot control his weight is unfit to hold a position of responsibility. Though Christie denied that these criticisms were at the root of his decision, he felt sufficiently stung by them to defend the idea that an overweight person can still be extremely accomplished in other areas.

Although such cultural shifts should not be dismissed lightly, there is also evidence to suggest that what is going on reflects a change in the nature of food choice itself. What we eat is beginning to be recognized as having an impact on others. First, the evolution of new market structures, as well as a more widely recognized appreciation of market dynamics, is tying food choice more firmly to the social justice issues that preoccupied the likes of Malthus, Mill, and Marx. Second, this same economic analysis is paired with trends in environmental philosophy, and the pairing has created an environmental ethics of food. Finally, prudential and existential concerns about health and safety have been reshaped by the logic of risk. Complex transitions are occurring in this third arena. On the one hand, dietetics has become a domain of personal vulnerability calling for regulatory action on moral grounds. What is vulnerable may be one's health, as in the case of food safety or nutrition, but it may equally be one's identity or solidarity with others as people attempt to achieve social justice and environmental goals through labels that promise "fair-trade" or "humanely raised" foods. On the other hand, practices that promote hospitable respect for personal dietary commitments or solidarity may run afoul of a philosophy of risk that emphasizes classic hazards to health and physical safety. All told, it begins to look less and less like food choice *can* be confined to the prudential realm.

The following provides a brief overview of the way that these three transitions in the nature of food choices are evolving. The three transitions should be regarded as heuristic rather than definitive. In fact, each category of change in the nature of food ethics blends into the other two. In the end, cultural transformations that make our current era seem more akin to the attitudes of the ancient world than to the high-water mark of the modern era's fascination with rationality enter the picture as well, though I will argue that we should broach the possibility of high food ethics with caution.

Food Ethics and Social Justice

As already noted, philosophers and social theorists have been probing the nexus of hunger, poverty, and distributive justice for some time. They have not used the term *food ethics* to describe this work. Nevertheless, the industrial revolution made access to food a key problem for social justice. As populations became less agrarian it was natural to see hunger as a social problem rather than a natural privation or an act of God. The famines of the distant past may have caused starvation and nutritional deprivation, but to the extent that famines were interpreted as naturally or divinely caused, the hunger that accompanied them was not viewed as a moral problem. But when the root cause of hunger can be understood to result from social or economic organization, or when the elimination of hunger can be seen as a clear technological prospect, it becomes possible to suggest that human beings have a moral responsibility to do something about it.

As Hub Zwart notes, the connection between food security and social justice can be discerned fairly clearly in nineteenth-century political economy. The thread continues into our own time, and quite a few distinguished philosophers have contributed to the analysis. Amartya Sen, the philosopher-economist, produced a clear analytical study proving to everyone's satisfaction that lack of food was not the root problem in any of the twentieth century's most notable famines. Thomas Nagel, Henry Shue, and Onora O'Neill are neo-Kantian philosophers who stress the interconnectedness of global economic systems. They assert that the wealth and comfort of middle-class Westerners is causally responsible for the deprivation of the poor in distant lands. Recently this theme has become a centerpiece in Thomas Pogge's approach to development ethics (though Pogge is not especially focused on food). In sum, where we once saw hunger and famine as needing our charity, the new ethics says we must now see that responding to such deprivation is a duty of justice—an ethically mandatory need to reform social institutions.[9]

A similar line of argument, but with an importantly different emphasis, has been put forward by the utilitarian philosophers Peter Singer and Peter Unger. Both noted that the suffering experienced by the hungry must by any standard be rated as of far greater moral significance than the satisfaction enjoyed by middle-class people who expend disposable income on fashion items, luxury goods, and personal entertainment. The utilitarian maxim—act so as to produce the greatest good for the greatest number—implies that one should give the money that would be spent on such goods to any of the worthy organizations fighting to address hunger in the developing world. Yet, of course, we routinely fail to do so. Singer and Unger consider the possibility that this failure is, in some sense, evidence *against* utilitarianism as an ethical theory, and it is this element in their work that has attracted most comment by other philosophers. But they both conclude that whether or not we can live up to the strictest interpretation of the utilitarian maxim, we are clearly obligated to be doing much more to relieve hunger than we currently are.[10] It is this component of their work that is crucial for food ethics.

Like the earlier analyses offered by Malthus, Mill, and Marx, these characterizations of hunger as a moral problem emphasize social and political organization and its impact on the needs of the poor. They do not tell us *what* we should eat. However, they do represent threats to the food security of others as a problem for which each of us should take some degree of personal responsibility. This is especially the case for the utilitarians, who argue that we should forego some of our consumption in order to facilitate consumption by those who have less. While not singling out the *food* consumption of comparatively better-off individuals, utilitarians argue that consumptive activity should be evaluated against an ethical norm of efficiency. Consumption of luxury goods (including luxury foods) is of comparatively little moral value, especially when it comes at the expense of high-value consumption needed to secure subsistence and nutritional health for the poor. The argument attacks *any* type of luxury consumption—buying designer clothing or expensive automobiles—but it also has clear implications for the ethics of eating a sumptuous meal at a high-end restaurant. Such reasoning might reasonably be applied to upper middle-class people who pay more for foods at boutique supermarkets. If we see our dietary choices as needing to contribute to an optimal amount of good in the world, we will see many of our meals imposing moral costs on the poor that simply cannot be justified.

And indeed a new set of market institutions is evolving. The food industry itself has worked to establish standards that align the developed world's pattern of consumption with improved livelihoods for poor farmers. These include food safety, purchasing, and sustainability agreements that largely work behind the scenes. They have emerged in part out of the food industry's response to its own ethical responsibilities and in part due to the market opportunity created by better-off consumers who feel the ethical pull. The most widely recognized of these efforts has been the development and promotion of the "fair-trade" label. Most successful and widespread among coffee producers, the label was created by ethically motivated purchasing co-operatives intended to insure that poor coffee growers receive a larger share of the price an end consumer pays for their beans (or their cup of joe, as the case may be). One problem with the fair-trade movement is that it may be becoming a victim of its own success. As the global market for high quality beans booms, conventional buyers are offering very competitive prices to coffee farmers, who in many cases are showing very little loyalty to the fair-trade cooperatives. It remains to be seen whether this model is economically sustainable. It is nevertheless clear that evaluating these standards and labels has generated a burgeoning literature of commentary in this particular corner of food ethics.

Some of the most visible commentary is frankly cynical. Slavoj Žižek has a video on YouTube in which he pillories the Starbucks Coffee franchise for attempting to enroll their customers in a form of "cultural capitalism" by purchasing fair-trade coffee.[11] The merits of Žižek's critique notwithstanding, the mere fact that a well-known philosopher has broken with five centuries of tradition in the discipline to express an opinion on whether or not we should drink one cup of coffee rather than another speaks to a point of more immediate relevance. First, food ethics is clearly back. Žižek's Starbuck's tirade is *not* a metaphor (or at least not only or even *primarily* a metaphor), nor is it even a more conventionally Marxist concern with the distribution of subsistence goods. He's talking about middle-class people drinking Starbuck's coffee and critiquing their reasons for doing so. However, both Žižek and the utilitarians are in fact less interested in food as such than in a more comprehensive concept of material consumption and its wider meanings. There is thus a sense in which they are continuing to conduct an older philosophical debate that has its roots in a time when no self-respecting philosopher would write (or make videos) about food.

It is also clear that standards and labels have brought social justice concerns into a shared economic space with environmental concerns. Labels that promise sustainability may be focused on workers' rights, or biodiversity, or both. To the extent that such schemes actually *do* further the goals they promise, we can begin to see our consumption of certain luxury commodities like tea and coffee as helping rather than harming the poor. So long as we take pains to ensure that our food purchases support small-scale farmers and peasant producers rather than large multinational corporations, we can feel that what we eat is ethically justified. Or if we take Žižek's perspective, we should resist this feeling, even if we participate in the practice. In either case, dietetics and personal choice enter the domain of social justice in a profound and potentially powerful way.

When the efficacy of fair-trade labels is contested, the thrust of the contestation lies in the social causality that links a consumer purchase to its social goal. Proving that the purchase of organic coffee from the co-op or coffee bar really does any good for Latin American peasants is a daunting task. The debate of social causality will be partly philosophical and partly empirical. Anyone who has followed an economists' discussion of market performance knows how complex the issues become, and how well-meaning initiatives can have effects that are quite contrary to what was intended. And if this social causality is so intricate, perhaps this fact reinforces Žižek's skepticism about the overall project of saving the world through better shopping. Although the details of such debates are deeply relevant to food ethics, we do not need to get so deep in the weeds right now. The larger point is simply that classic debates in political economy have now been plausibly linked to dietary practice. It is one example of why food ethics is with us again.

Food Ethics and the Environment

Some authors mark the beginnings of contemporary environmental consciousness with the publication of Rachel Carson's *Silent Spring* in 1962, while others see it in a longer-term historical perspective that stretches back at least to George Perkins Marsh's book *Man and Nature* or perhaps even German romanticism and the noble savages of Jean-Jacques Rousseau. However one styles the rise of environmental ethics, food choices are increasingly coming to be seen as morally significant because of their impact on nonhuman others. Ethical vegetarianism is the most obvious example. Though philosophical arguments against eating meat that rest on the fact that a carnivorous diet requires the death of an animal are as old as philosophy itself, it is patently obvious to anyone having frequent interaction with academic philosophers that something novel is afoot. Ethical arguments for vegetarianism are proliferating rapidly and the proportion of people finding them persuasive is growing markedly.

Medical vegetarianism understands dietary choice in prudential terms. If I choose not to eat meat because I'm concerned about my health, then at bottom it's all about me. These medical motivations may have been preceded historically by religious concerns. Religious ideas sometimes build an unstable bridge between prudential self-interest and morality, and religious vegetarianism provides an example. Some health-oriented vegetarians align their personal health to teleology (a divine or evolutionary plan for harmony among living beings). As Tristram Stuart documents, there are explicit statements of such teleology that reinforce a religious obligation to elect a vegetarian diet, and that see ill health as God's punishment of the wicked people who neglect this duty. In aggressively evangelical cultures, the compulsion to enroll nonbelievers elevates all religiously based standards to the status of moral norms: *everyone* must obey what the believers take to be God's will. Some of the most ardent vegetarians of the seventeenth century were evangelical Christians who believed they had a duty to spread their version of God's word.[12] However, principles of religious tolerance dictate a different reading of religious duty. Believers are understood to accept a religious obligation to elect a vegetarian diet, but those who are not members of the faith would not be thought unethical for eating differently. Hindu vegetarianism is, on such grounds, comparable to Muslim or Jewish proscriptions against eating pork. And of course one could decide to become a vegetarian purely out of personal taste.

In contrast to all these rationales for vegetarianism, ethical vegetarians understand themselves to have chosen against eating animal products out of respect for the interests of the animals themselves. The philosophical basis for such respect is articulated in various ways. In a formulation offered by Tom Regan, nonhuman animals with sufficiently sophisticated neurology are understood to possess a form of subjectivity or inner life. On Regan's view, our moral obligations to all animals, including other human beings, arise out of a duty to show respect for the subjectivity and interests associated with this mental activity. Minimally, we should not sacrifice these interests simply to satisfy a comparatively trivial interest of our own. As long as nutritional and subsistence needs can be met without eating animal protein, consumption of meat and other animal products (milk and eggs) should be regarded as a relatively insignificant interest, especially in comparison to the animal's interest in living out its life and avoiding pain or confinement. A meat-free diet is therefore incumbent on all human beings, excepting perhaps those few impoverished people who could not survive without consuming animal products.[13]

Peter Singer has formulated a less onerous ethical vegetarianism. Singer argues that the ability to experience pain and suffering is the ethical basis for our obligations to other animals. We are not permitted to inflict pain unless the reduction of even greater pain is expected to be the result. Given the reasonable presumption that human beings have a far more sophisticated ability to experience both satisfaction and suffering from events that affect others (especially loved ones and family members) as well as to experience psychological well-being associated with future expectations, events that involve the death of a human being are typically associated with incomparably greater suffering than the deaths of most nonhuman animals. As such, the painless slaughtering of livestock is much less morally significant than the suffering that food animals endure while they are alive. Although Singer has argued for vegetarianism as a form of protest and as practical dietary ethics, his view would permit the consumption of animals raised and slaughtered under humane conditions. His philosophy requires vegetarianism given the suffering that he believes livestock endure under the conditions typical of industrial animal agriculture.[14]

Both of these approaches retain the modern convention of treating wholly self-regarding acts as nonmoral. Dietary choices are moralized because we have become aware of and sensitive to their effects on nonhuman others. Additional forms of environmental consciousness can also extend the scope of our other-regarding duties. Francis More Lappé's book *Diet for a Small Planet* advocated a reduced meat diet on grounds wholly distinct from those of Regan and Singer. Lappé notes that animal diets are an ecologically inefficient source of nutrients required for a healthy human diet. Poultry must consume roughly twice as much plant protein in their feed to make a pound of animal protein available for human consumption. Cattle may consume four times as much. Since humans can survive on either plant *or* animal protein, it would be much more efficient for humans to simply eat the grain that is fed to animals during livestock production. This would reduce the amount of land needed for agricultural production and return a significant portion of our planet to natural ecosystems.[15]

Lappé links dietetics to a broadly comprehensive environmentalism that encompasses duties to wild nature as well as to future generations. Although the original version of her argument mounted in the 1970s neglected the fact that large herbivores such as cattle can graze on grasses that humans cannot consume, recent scientific studies suggest that there is also a climate ethics dimension to her basic argument. In addition to feed inefficiencies, livestock emit greenhouse gases as a natural byproduct of digestion, flatulence, and defecation. Ruminants metabolizing grass are especially serious offenders, so although range-fed beef and milk production might elude the force of Lappé's original efficiency argument, they are deeply problematic from a climate perspective. The Food and Agricultural Organization of the United Nations has estimated that as much as 20 percent of global greenhouse gas emission may occur in connection with livestock production.[16] Thus, to the extent that mitigating the release of greenhouse gases into the atmosphere is viewed as a moral obligation, humanity has an even more powerful rationale for limiting animal protein consumption.

There are, of course, many details in these environmentalist arguments that can be contested. How are we to compare the subjective experience of humans and nonhuman animals? Does it matter whether we base our environmentalism on a concern for future generations, on duties to other sentient creatures, or on some more expansive understanding of the intrinsic value found in natural

ecosystems? These questions lead into key debates that have occu-
pied environmental philosophers for the last four decades. An
alternate line of questioning might probe the way that different
systems of food production affect animal suffering, biodiversity,
climate change, or ecosystem processes. Simon Fairlie has assembled bits and pieces of environmental data into an argument for
what he calls "the default meat diet." It eliminates meat from animals that have been fed grain raised specifically for animal production, but encourages continued production (and consumption) of
animals pastured on grass or fed otherwise unusable food wastes.[17]
Here philosophy and science become intertwined as one attempts
to ferret out matters such as the energy consumed in transporting
a boatload of grain across the ocean as compared to driving numerous trucks and automobiles from farm to market in order to supply a local food system. Again one finds oneself down in the weeds
of food ethics, trying to sort out the respective costs and benefits
of alternative systems. At this juncture, it is important simply
to point out that what was once regarded as a purely prudential
domain is now seen as fraught with moral significance. Those of us
who try to answer such questions might differ with Regan, Singer,
or Lappé about the basis or implications of our dietary choices, but
the very structure of our debate reveals that we have accepted the
moralization of dietetics in a fundamental way.

Food Ethics and Risk

Each of the philosophical developments chronicled above understands the ethics of food choice as a function of the way that
dietary habits affect others. In the case of environmental ethics it
may be nonhuman others, or it may be generations of humans yet
to come. The moral status of nonhumans and future generations
is frequently debated in contemporary philosophy, but no matter
which side one takes, the debates themselves illustrate why food
choice has again become a moralized domain. In the case of social
justice, our increasingly sophisticated understanding of trade and
economics has given rise to the view that many of our routine food

choices actually *do* affect unseen human beings in distant locales. Schoolchildren from at least one generation were told to eat everything on their plate because "people are starving in China." Quite rationally, they grew up wondering how eating their broccoli or green beans could possibly do anyone in China any good. The theorists of social justice have now filled in this gap in our understanding, though not, perhaps, in a way that supports this particular bit of parental advice. It is not literally the *eating* of food that affects others, but the way that dietary choices are converted into political acts in virtue of the purchase that precedes the meal. The political economy of these choices can be debated, but the fact that we are debating shows how diet has become a moral issue.

Something different is going on when we turn to questions of risk. Part of the reason that "You are what you eat" succeeds aphoristically is that on a literal reading it seems to state a simple truth. Our bodies are materially composed of water, minerals, proteins, fats, and other chemicals that we have ingested as food and drink. One could object that we are also composed in small measure of substances that we have absorbed through our pores or by breathing, but this is slicing the carrot in a deliberately obtuse fashion. "You are what eat" is pretty much true on a biophysical level. One could quantify the percentages and work out the scientific details if one were so inclined. Indeed, molecular tests are now available that can show what percentage of various plant proteins went into the construction not only of our own bodies but also of the animal bodies we may have consumed as meat.

The fact that food is taken into and becomes one's body is certainly among the reasons that some ethical vegetarians feel as strongly about their diets as they do. The sense of revulsion people feel on learning that they have ingested some proscribed or filthy substance, even when this information is received well after a meal has been consumed, is not unique to vegetarians. Mary Midgely has argued that such feelings of revulsion provide a sufficient reason to refuse to eat genetically modified (GM) food.[18] In the case of both vegetarian and anti-GM dietetics, these strong emotional associations pair up with the judgment that food consumption is causing harm to nonhuman others. Public opinion research indicates that most people who object to eating GM foods do not take the foods to be unhealthy for themselves but rather believe that production of GM crops harms the environment.[19] In connection with genetic engineering I myself have argued that there is one important sense in which concern for what we eat has moral implications: it is morally wrong to be coercive or deceptive about the food we offer to others. Failing to respect a vegetarian's preference is as much a moral affront as failing to respect a religious Jew keeping kosher. If someone is repulsed by the very idea of genetically engineered food, it is disrespectful to place him or her in a position where they will eat such food without their knowledge, just as it is disrespectful to structure the food system so that it is impossible for them to avoid it. In such cases, respect for dietary choices is a way of respecting the cultural and symbolic identity that the person has chosen. Denying others the right to construct and live out such an identity is to deprive them of their dignity. It is to treat them as something less than a morally significant other.[20]

Perhaps one will say that it is just the coercion and deception that is the basis of the harm here, in which case the connection to diet is purely incidental. Yet there are reasons to think that coercing or deceiving people about their food represents something of a special case. For starters, misleading people about their food straddles a cultural chasm. On the one hand, fooling people into trying something new is widespread, and there seems to be something of a license to pull the wool over the eyes of notoriously picky eaters: "I knew you would like it if I could just get you to try it!" On the other hand, mischievous disregard for expressed food preferences can have dire consequences when food allergies are involved. What is more, people are increasingly developing food preferences and beliefs that reflect a lack of confidence in the industrial food system, or a distrust of the science used to determine whether a given food is safe. People in the food industry are routinely dismissive of such beliefs, and anyone who interacts with food scientists will hear such attitudes described as irrational. Yet most of us have ingested substances once thought safe but now known to pose risks. They include banned pesticides, residues from plastic packaging, food colorings or additives, and drugs administered to food animals. We carry the residue of some of those substances in our body tissues today. We are what we ate, and we may yet pay for it.

In the face of such considerations, it is difficult deny the possibility that we are continuing to expose ourselves to unknown dietary hazards. For myself, I am more than willing to place my bets with the toxicologists and microbiologists who study food safety from a scientific perspective, but in saying this I acknowledge that I *am* making a bet with my personal health. I would not deny other people the right to formulate their own judgments about the risks that they will and will not take. More broadly, to place roadblocks in the way of someone's ability to select which risks they want to take is an affront to their dignity and autonomy, whatever the current science says about the safety of a given food. To be sure, not everyone wants to be burdened by making risk calculations whenever they sit down to some steaming mashed potatoes. It is reasonable to align public policies with the best scientific consensus. But there are numerous areas where people are quite prepared to take risks that the current scientific consensus does not support. Drinking raw, unpasteurized milk would be a case in point.

And then there are cases where risks that have been judged negligible by science seem either too large or unnecessary to given individuals (including some scientists, I should note): pesticide residues, preservatives, coloring agents, and packaging ingredients such as BPA (bisphenol A). Why shouldn't people be allowed to take their own precautions in such instances? There is thus a perspectival or positional component to the ethics of food safety. I want my food choices to align with the best contemporary science, and I am glad that our food safety policies are structured in accordance with it. I can agree that aligning with a scientific assessment of risks and benefits is "only rational" when viewed from my own perspective. On the other hand, I see no particular reason why my perspective should determine the food choices of others. If someone else thinks that tradition or religion is a more reliable source of dietary advice, imposing my own science-based perspective upon them becomes ethically problematic. I am therefore opposed to the claim that what seems "only rational" to me is in fact the only rational perspective for anyone to take. Equating scientific rationality with moral rationality in the matter of food safety risks is both unwarranted and potentially unethical.

A brief story told to me by a colleague in the agricultural sciences illustrates the point. He grew up on a farm in the Dakotas and, like many in generations past, was raised on milk from the family cow. It seems that a city-bred priest came to spend a few weeks with the family on their remote prairie farm. Horrified by the sight of milk coming from barn to table in an open pan, the priest insisted that the farmer (my colleague's father) make a thirty-mile trek to the closest market so that milk could be served as it should be: from a milk carton. The farmer, being wise, did so without protest. But when the milk carton was empty, he brought fresh milk from the cow into the kitchen in a pan, and my colleague's mother poured it into the empty carton and deposited it in the refrigerator. The priest continued to enjoy his daily milk, commenting on its wonderful flavor. This practice continued for the duration of the priest's visit. He never discovered the deception, and to all appearances he was none the worse for it.

I am inclined to laugh at this story in the same way that many rural folk traditionally would: the goofy priest with his crazy ideas is cannily taken in by the clever farmer. Yet it is clear that this innocent deception disregarded the priest's wishes, and in that sense, at least, the farmer's treatment of the priest is hardly consistent with the ethical norms of respectful hospitality. There are plenty of weeds to wander into here as well. The philosophy just outlined suggests that our public policy should allow for a wide variety of personal dietetic codes, but what about the nutcase who *wants* to eat a bowl of cornflakes laced with toxic pesticide residues? Many food scientists would say that something very much like this situation is actually not as improbable as it sounds. Without careful controls, organic or raw foods may carry a much higher load of dangerous bacteria than products of the industrial food system. In fact, although the farmer broke no laws by serving raw milk in his home, unpasteurized milk is viewed as a significant hazard by food safety experts, and in many locales it is unlawful to sell it or to serve it for public consumption. The details of these policy questions provide yet another avenue into more nuanced food ethics. More generally, it seems that when it comes to risk it is the question of what we choose NOT to eat that creates an opportunity to frame dietary questions in ethical terms.

Food Ethics and Cultural Identity

In comparison with many other consumer preferences, food choices are significant in virtue of the way that foods are employed in ritual and symbolic activities, as well as the fact that foods are taken into our bodies. Anthony Giddens has suggested that consumption practices "give material form to a particular narrative of self-identity."[21] Telling ourselves certain stories and then living them out through our material practice are complementary activities that produce a more meaningful life. The narratives perform quite a number of functions that followers of John Stuart Mill may have a tendency to overlook. The stories or myths that animate our daily practice serve as a common source of memory and experience. People living in community (or even just proximity) draw upon this pool of experience in learning how to get along with one another. The stories we tell about our food become a way to become more self-aware about our material life, about the day-to-day doings that usually go unnoticed but actually occupy a significant portion of our time. These stories allow us to converse about what we eat. The conversation in turn allows us to explore the similarities and differences among diverse family and personal ways of living. Foodways then supply us with metaphors that enrich our moral imagination. Performing the rituals of preparing and communally enjoying our food (e.g., cooking, serving, and then eating) makes the symbolic forms—the stories and metaphors—real. This give and take between symbolic projection and material performance instantiates what we mean by an ethic in the broadest and deepest sense.

A narrative that celebrates the aphorism "You are what you eat" may be singling out the material practices of food consumption as a particularly important example of "performative intentions" or norms and values that we hope and expect to enact in our day-to-day activities. The arguments summarized so far support imperatives of respecting others through our dietary practice, and for respecting the dietary practice of others. Do they also provide a basis for turning an ethically reflective gaze onto our own diet as a symbolic performance of cultural identity? In *The Agrarian Vision* I have drawn upon Albert Borgmann's work on "the culture of the table" to address this question. For Borgmann, the nexus of producing, preparing, and consuming food is a "focal practice" that can generate communal projects and shared meanings. Borgmann advances his analysis of focal practices in response to an observation that is central to his overall philosophy: as contemporary life has become more and more characterized by ready, efficient access to goods, the unintended consequence is that our lives have become emptied of meaning.

The basic idea is that once upon a time (and actually not so long ago) our daily lives were absorbed with onerous tasks: cutting wood for the fire, *tending* a fire for heat or cooking, growing vegetables in our own garden, helping a neighbor raise a barn, or chatting with that neighbor to get the latest news from town. We can be thankful that industrial technologies make all of these tasks less physically taxing, and that as a result we have more time for other pursuits. But as goods such as fuel, shelter, food, and information are delivered rapidly and without the annoyance of time-consuming involvement, we are discovering that those annoying delays had an unnoticed side-benefit. Although people were often exerting a great deal of physical energy, this work was also a source of enrichment, fellow-feeling, and community. People felt a sense of accomplishment when the work was done, and it may have been keeping them physically fit, as well. What is more, the work often connected people. Sometimes it was literally done in a group, but even when it wasn't, there were complementarities among the daily tasks that made people aware of their interdependence on and interaction with one another. Borgmann is not suggesting that we go back to the old ways, but we do need to find new life-enriching and mutually engaging ways to spend our time. He calls these "focal practices" because they give our lives a sense of being centered. Our preoccupation with them yields indirect or unexpected sources of meaning and conviviality.[22]

But while the production, preparation, and consumption of food can indeed be an important source of symbolic interaction for us, Borgmann mentions the culture of the table as but one of *many* focal practices that can do this. It is important to recognize that law, policy, and technological practice can each undercut existing focal practices. Borgmann argues (and I agree) that protecting the space in which focal practices can be performed is a crucial and underappreciated theme for ethics in contemporary culture. The arguments already sketched *do* provide a basis for recognizing the right to engage in performative identity construction up to but not beyond the point that doing so compromises the ability of others to do exactly the same thing. This is what Mill's doctrine of liberty hoped to achieve. We *should* engage in these practices, while *also* being sensitive to the effect that our focal practices might have on others. There is no issue with feeling *entitled* to tell food stories, nor with enacting them, so long as doing so does not compromise the rights of others. But we can question whether people (we ourselves, but also others) are ethically *obligated* to cultivate a performative food ethics around cultural identity construction.

And my answer is, "No, you are *not* what you eat." I have surveyed a number of ways in which the present-day global food system connects our dietary choices to effects on others that are worthy of ethical reflection or debate. We are embedded in social relations that cause exploitation far from the site at which our meals are consumed. The ethics of food security receive more focused discussion in several of the chapters that follow. The impact of our food choices on the environment can be interpreted either as harm to nonhuman others or to unborn generations (Environmental food ethics is the focus of Chapter 6). Finally, the tortuous logic of risk can place us in situations where norms of respect or hospitality mandate our accommodation of perspectives quite unlike our own. The risk issues in the debate over genetic engineering of food provide an extended example that is discussed in Chapter 8. But these topics in food ethics do not require that we regard dietetics as deeply constitutive of moral character or cultural identity in quite the manner suggested by ancient concerns with moderation or temperance.

Food Ethics and Liberal Society

Health risks aside, markets usually mediate the material practices that are most likely to have an impact on someone else's rights. It is not literally what we eat that we should be concerned with, but what we buy. This provides the basis for qualifying the sense in which "You are what you eat" can be understood as an ethical claim. Consider, for example, that I am enjoying a snack of fair-trade coffee and free-range eggs.[23] A critic accosts me for the suffering that coffee planters inflict on peasant workers. She then berates me for the pain inflicted on caged hens, concluding with the aphorism, "After all, you are what you eat!" My comeback is obvious: "You are mistaken about the facts. If I am what I eat, I am in this case someone helping these causes rather than harming them." I can make this claim not because I am drinking coffee or eating deviled eggs but because I have paid premium prices for these goods to support social and animal justice. Evaluated in cold logic, a vegetarian's refusal of meat on the plate does about as much good for the animal in question as our schoolchild's consumption of broccoli does for the starving people in China. The effect of diet on others is thoroughly mediated by market structure. Once the purchase is made, there is a sense in which the moral impact is complete.[24] Eating neither adds nor subtracts.

Now there is clearly a close connection between the food we purchase (or that is purchased for us) and the food we end up eating. So it is perfectly reasonable to articulate our food ethic in terms of dietetic rules. Nevertheless, the moral potency of these rules is largely exhausted once we have signaled our intentions through market channels, at least insofar as we are concerned with the impact of our conduct on others. One demonstration of this point resides in the way that with different economic arrangements, virtually all of the things we should not eat become things that we *should* eat in order to reorient material practices in more just, sustainable directions. Ethical vegetarians, for example, might help improve animal welfare by communicating a willingness to pay for animal products produced under humane conditions. And as we shall see in Chapter 5, this need not be restricted to those vegetarians who see suffering, rather than slaughter, as the ethical problem.

What about putatively repulsive or risky foods? Is avoiding them a form of ethically laden identity construction? Again, anyone has a *right* to adjust diet according to tastes and beliefs about risk, just as they are within their rights to affiliate with a political party or a religious faith. But just as they have no *obligation* to affiliate with any particular religious faith, they have no moral duty to adopt any particular dietary rules. People who choose to remain secular (about food *or* faith) may be performing some broad ethical narrative in Giddens's sense, but this does not imply that they have engaged (or should engage) in a deliberative, reflective philosophical review of the reasons for doing so. Once people have established a particular food practice, such as drinking unpasteurized milk or eating organic food, they may feel themselves to be performing a social identity when they take nourishment in this particular way. They may experience this as an *ethical* performance, as something they are morally obligated to do (but then again, they may not). This feeling of ethical significance can certainly be associated with foods believed to be risky, as when someone eats organic avocadoes in order to avoid exposure to pesticide residues. For a parent who feeds their child organic avocadoes the feeling of ethical significance is even more profound. Would they also be justified in a kind of evangelism for their view of what people should eat? In answer it seems (once again) that evangelism of any kind would be protected by freedom of speech, but the important questions here concern whether or not the putative risks are borne out by the facts.

One might draw upon other bits of philosophical terminology to clarify what is at stake. John Rawls draws a distinction between matters of justice, on which we must reach some form of structured agreement, and conceptions of the good, which must be plural in liberal societies so that individuals are free to pursue different plans of life. Tolerating diversity in peoples' conceptions of the good life is, for Rawls, the central theme of political liberalism. Jürgen Habermas adapts a Hegelian distinction between *moral* discourse, which is intended to establish universal principles that everyone must live by, and *ethical* discourse, which communities undertake to establish shared principles of cultural identity.[25] Related to Habermas's usage, there are some feminists who use the term *politics* when they want to talk about the formation and enforcement of other-regarding norms, and reserve the word *ethics* specifically for the kinds of identity-building discourse that I have said we have no special ethical obligation to engage in.[26]

These are obviously not equivalent bits of philosophical terminology. The challenge of navigating inconsistent word use confuses everyone who takes an interest in food ethics. Once one gets beyond the ambiguous terminology, the main point to notice in the domain of risk-oriented food ethics is that liberalism cuts two ways. On the one hand, people are free to accept or reject science in a liberal society, at least in so far as doing so affects only themselves. On the other hand, a restaurateur is *not* at liberty to apply idiosyncratic theories of sanitation or food safety to kitchen hygiene precisely because such practices could harm others. In the domain of food, science is not something that can be culturally rejected in favor of antinomian food beliefs, because science represents the basis of our cultural standards for sanitation and public health. And public health *is* a matter of justice in a liberal state. To put the point in yet another way, people in a liberal society have a right to pursue their cultural identity through dietary choice even if it means rejecting science, but they do not have a right to impose a cultural identity that rejects science on others.

At this juncture, food ethics confronts Michel Foucault's powerful analysis of the links between social power and academic disciplines of knowledge, the practices of truth that are institutionalized and ritually performed in the various branches of the natural and social sciences. Foucault noted that knowledge disciplines such as sociology and psychology arose in response to a perceived need to govern and rationalize violence visited on the bodies of political subjects. While he was most intent on exposing the way in which disciplines arose to normalize certain sexual practices, he did not fail to note that power was also applied to bodies with respect to food and drink. A certain strain of lawless excess showed itself in acts or accusations of cannibalism, of monsters who flouted every sense of propriety. In rebellion against propriety, some even forced others into acts of anthropophagic excess. Such acts challenged conventions that undergird a social order defined through *what* one eats as much as *with whom* one enjoys sexual intercourse.[27]

As will become clearer at several junctures in the book, there are many ways in which our food habits define who we are, and one need not resort to such extremes as anthropophagi to illustrate them. Dogs, cats, and primates are currently eaten in various parts of the world, but some people who give little thought to eating a cow or a pig will regard the consumption of dog meat with unspeakable horror. And to shift from the sublime to the ridiculous, one can shock cultural sensibilities and spoil a friendship simply by serving a commonly recognized food item at the wrong time or place. Try putting the salad croutons on top of someone's chocolate sundae and you will see what I mean. There thus clearly *is* a sense in which culturally significant foodways take on a normative significance that goes well beyond the potential for bodily harm. I am quite willing to classify the offense that would be taken by an observer of religious or cultural dietary norms as a form of harm. If someone decides to serve up a plate of roasted dog or cat at a community picnic, it will matter a great deal what community is picnicking. In virtually all Western societies, someone lured into eating such a dish could legitimately claim to have been harmed.

This means that there is a conversation, perhaps even a negotiation, that begins to pull the practices of dietary identity formation more squarely into the moral sphere.

Challenging Liberal Food Ethics

Conversations like this are already underway. For example, Kari Norgaard, Ron Reed, and Carolina Van Horn describe the history of displacement and unjust treatment of the Karak tribe in California. Land to which the Karak were entitled by treaty and historical occupancy was transferred to the US Forest Service (USFS) over a century ago. Today the tribe is not able to eat the diet of meat and fish that was characteristic of their ancestors because USFS policies deny them ready access to hunting and fishing. Instead they must go to the store and purchase food. Some tribal members have claimed that they suffer from nutritional deficiencies and food insecurity as a result. Norgaard, Reed, and Van Horn quote tribal member David Atwood, who says, "Our way of life has been taken away from us. We can no longer gather the food that we [once] gathered. We have pretty much lost the ability to gather those foods and to manage the land the way our ancestors managed the land."[28]

How does this case exhibit food ethics? On the one hand, the historical acts of taking tribal lands and confining tribal members on reservations qualify unambiguously as ethical issues. Even someone who seeks to justify the actions that Norgaard, Reed, and Van Horn condemn must recognize that they will have to produce an argument to do so. Similarly, the matter of what should be done about poverty, unequal access to resources, or discriminatory enforcement of laws that regulate hunting and fishing are ethical issues. The allegation in each instance is that individuals have been harmed by other individuals, by the government, or by distributive inequities built into the structure of society. These are the kinds of problems that liberal societies were in some sense designed to address. Liberal approaches to ethics are less facile in addressing questions where structural or institutionalized racism is at work, where law, policy, and customary norms continually reproduce inequalities along racial lines. Yet here too, the "other-regarding" quality of structural racism—how routines practiced by privileged groups affect others—is incontestable.

But Norgaard, Reed, and Van Horn also write that a traditional Karak diet would have averaged something on the order of 450 pounds of salmon per year. Today the average Karak eats more like 5 pounds of fish a year. Should today's Karak understand the performance of their cultural identity in terms of restoring this dietary practice? And if they do interpret their present-day identity in terms of this long-past historical norm, will restorative justice (discussed in Chapter 2) provide the basis for a claim that they can levy against the rest of society? Must the rest of us ensure that the Karak are able to eat as much salmon as they want? At several points it seems like this *is* what Norgaard, Reed, and Van Horn associate with nutritional justice. One problem arises immediately in the wake of interpreting the ethics of performative dietetics in this way: is *anyone* entitled to select some dietary practice of their ancestors as their norm for a present-day cultural identity? The classic answer that John Stuart Mill or John Rawls might have given for such a question is that anyone is entitled to aspire to whatever conception of the good life they want, but aspiration does not provide a basis for making enforceable claims on anyone else. To be sure, the past injustices cited by Norgaard, Reed, and Van Horn make the case of the Karak different from someone's purely arbitrary commitment to ancestral foodways. Nevertheless, this case exemplifies how sticky problems in the assertion of food-based cultural identity claims can become.

I thus lay down the following challenge: many, if not all, societies have celebrated foodways as a component of identity, while relatively few have incorporated practices of diversity and tolerance. The moral narrative that stresses religious tolerance, cultural diversity, and individual freedom has become far more central to the self-understanding of Western culture than our food preferences. There is a lot of work to do in reining back the excesses of

Western culture when it comes to race, gender, and ethnic identities, but I see no reason to undo the commitments to tolerance and diversity. Adapting a line of argument put forward by Volkert Beekman, we should recognize that dietetics represents an important source of meaning and self-identity for ourselves and others, but we should not think of it as having the kind of force that would justify any particular set of foodways as a binding social norm. Beekman argues that we should not put the power of government behind the advocacy of dietetics.[29] Dietetic rules have moral content when they concern the *right* to engage in focal practices that are meaning or identity building, and when they have socially mediated *effects* on others. I am saying that we should take care not to let our enthusiasm for focal practice bleed over into something that could become repressive for others.

Conclusion

In summary, there is indeed a place for a revival of food ethics, and we should expect that philosophers will find more frequent opportunities to reflect upon food production and consumption. This is not, however, a reversion to an earlier era in which the practices of self-discipline and diet are taken to be a mark of moral character and personal virtue. Instead, we have at least three domains in which the organization of our food system presents us with moral problems. First, industrialization and the emergence of post-industrial societies have created dense networks of social causality affecting the worldwide production and consumption of food. Food ethics will continue to be an ever-increasing component of social justice. Second, there is the environment. The fact that we will continue to debate both the moral significance of environmental impact and scientific basis for tracing it underscores the way in which even those who reject the claims of ethical vegetarianism cannot escape the need for food ethics. Finally there is risk not only to health and bodily integrity but also to traditions, practices, and forms of social solidarity.

I have suggested that the primary focus of food ethics should lie in the way that our social institutions, market structures, and public policies entangle us in other-affecting consequences. Sometimes it is we who are affecting others, as when we purchase food products that stimulate ill treatment of other people or nonhumans. Sometimes it is others who are affecting us, as when we bear food risks that we have not chosen to take. The literal concern with what we actually eat could be thought of as a prudential, aesthetic, cultural, or possibly religious concern. Philosophical considerations would naturally arise in negotiating such concerns, but I am drawing a line—a vague and broken line, to be sure—between these domains and ethics. Nevertheless, in each of the chapters that follow, issues of meaning or cultural significance emerge *along with* the impact of our myriad food practices on others. This only goes to show that sometimes drawing the line that separates the ethical from aesthetic and cultural domains actually *is* a task that belongs properly to ethics. It proves that food ethics is a philosophically contestable domain of inquiry.

Chapter 2

Food Ethics and Social Injustice

Many of the food-system activists I meet are focused on the environmental impact of farming and food production, but almost as many are appalled by the fact that laborers in the industrial food system are among the most abused and exploited workers in the industrialized world. When they hear the phrase "food ethics," those who are less fully engaged in social movements for food-system reform will immediately think of community food pantries, soup kitchens, and other activities intended to feed impoverished and often homeless people. Although later chapters examine issues of hunger and food security through the lens of global development, social justice closer to home is the big thing that many would associate with food ethics. It is a primary motivator for many young people who become involved in food-system issues. Many key food topics can be approached powerfully through a political or philosophical ethic that builds on an analysis of poverty or of racial and gender oppression. This kind of philosophy examines the relationship between the history of equality and the inequalities that continue into the present age. These themes cannot be ignored in a book on food ethics.

Injustice in the Food System

The CBS television network broadcast a documentary on migrant farm labor over the Thanksgiving weekend of 1960. *Harvest of Shame* was based on reporting by Edward P. Morgan and David Lowe and narrated by the highly respected broadcast journalist

Edward R. Murrow. The film documented the poor living conditions and shabby treatment of workers who brought in the fruit and vegetable harvests on American farms. Interviews with white, black, and Hispanic migrants documented the workers' lack of access to basic necessities, such as water and toilet facilities, and vulnerability to arbitrary acts of nonpayment by farm owners, as well as the plight of children, who also labored, or who moved from school to school as migrants followed the harvest. Migrant laborers also suffered from the structural injustice of exploitatively low wage rates.. Viewers were equally shocked by interviews with seemingly uncaring farm employers, one of whom described the migrants as "the happiest people on earth."

Harvest of Shame brought the unfair treatment of farm laborers to the attention of American citizens enjoying the unprecedented economic growth and prosperity of the 1950s. The bitter irony of a well-fed television audience enjoying a feasting holiday at the expense of people trapped in perpetual poverty had a profound effect. The working conditions and poverty of farm laborers were virtually unimaginable to members of the industrial working class. The airing of *Harvest of Shame* led to a round of efforts at political reform. Yet while field laborers did achieve the legal right to minimal amenities (such as access to portable toilets), they continued to occupy the lowest rung on the ladder of social advancement. Farm workers lived in substandard housing, often without heat or running water, throughout the 1960s, and farm employers were exempted from federal minimum wage requirements. Only a decade after *Harvest of Shame*, Cesar Chavez's struggle to organize California farm workers again dominated headlines.

Although farm workers have enjoyed some intermittent episodes of relative progress during the half century that has passed since Murrow's reporting and Chavez's activism, exploitation continues. Barry Esterbrook's 2012 book *Tomatoland* documents human trafficking by labor contractors for large Florida fruit and vegetable firms, as well as the illegal and unethical practices of growers in the tomato industry who tried to repress the Coalition of

Immokalee Workers—a local organization attempting to improve working conditions. Esterbrook reports on labor contractors who lure recent immigrants into the United States with promises of food, shelter, and employment. Instead, these people wind up imprisoned in windowless buildings without bathrooms or kitchens. They are trucked to the fields in the cargo bay of open vehicles lacking seats, safety equipment, or protection from the elements. On top of all this, their earnings are garnished to pay for "housing and transport benefits." Many are not legally approved to work in the United States, and this makes them both fearful and willing to endure the confinement, beatings, and rape.[1] Whatever one thinks of their immigration status, the treatment these workers receive is a moral outrage. The entrapment and confinement that Esterbrook describes is unlawful and violates any reasonable standard of human decency. The willingness of growers and the food industry to turn a blind eye to these practices reveals a lack of basic ethics, even if the arrangements they have with labor contractors shield them from legal culpability. Exposés like those of Murrow or Esterbrook inevitably raise a further ethical question: can those of us who enjoy the benefits of these morally tainted practices, practices that seem to be endemic in the food system, hold ourselves blameless?

Well beyond the farm fields where workers must still cope with substandard working conditions, people employed in the food sector continue to exist on poverty-level and barely above poverty-level incomes. And while injustice in the fields may seem to have been with us always, in other sectors of the food industry things have been getting worse. Meatpackers once had one of the strongest unions and enjoyed a standard of living comparable to that of well-paid skilled labor in manufacturing and the automobile industry. Over the last thirty years, the power of these unions has been broken through the introduction of powered saws and knives that permit packing companies to employ unskilled labor. While conditions for workers in slaughterhouses and packing plants are better than those for Florida tomato pickers, these are now high-turnover jobs paying little more than the minimum wage. The work is dangerous and workers suffer from high rates of repetitive stress injuries. If they complain, they will very likely be fired.[2] Food industry firms in the retail sector make widespread use of part-time jobs in order to avoid paying benefits. Wages in restaurants and food service facilities also often hover near federal or state minimums. Although the situation of workers in Europe is more stable, in the Americas, income for wait staff, kitchen help, and dishwashers may depend on unreliable gratuities, and customers may deploy the power that comes from this situation to add an insult to the injury employees already suffer in the form of low and unreliable incomes.

Most significantly, employees on farms and in food industry firms may be especially vulnerable to more extreme abuses due to non-enforcement of existing laws. Esterbrook's reporting on migrants is but one particularly extreme example of failure to enforce the law. Restaurants and food industry firms are also cited for labor violations on a routine basis. A 2009 report found that roughly 23 percent of grocery stores and 18 percent of restaurants in the United States violate labor or employment regulations.[3]

These circumstances are not unique to the United States. Landless laborers exist throughout the world in both developed and developing countries. In 2014, the *Los Angeles Times* ran stories on even more dramatic abuses in Mexican fields where tomatos are grown for export markets. In non-industrialized countries, agricultural laborers are almost always the poorest of the poor, living under conditions that the World Bank defines as "extreme poverty"—an annual average income of less than one Euro per day. They perform seasonal tasks such as hand weeding and hand harvesting and may endure long periods of idleness when their labor is not needed by farmers who themselves may be only marginally better off in terms of income. Poor farmers may be vulnerable to extremely volatile markets and are often in the position of needing to sell their crops when market prices are at their lowest. Both farmers and farm laborers may then need to buy food in the off-season when prices are highest. Raj Patel's book *Stuffed and Starved* portrays episodes of abuse and deprivation in Europe as well as in rapidly developing areas such as India and Pacific Asia. His title recalls the deep irony of a world in which farm and food workers suffer from malnutrition and poverty, while wealthy people in the industrial world increasingly suffer from the diseases of affluence—diabetes, heart disease, and other ailments of overconsumption.[4]

Injustice as a Philosophical Problem

The plight of the exploited human beings who do the day-to-day work that delivers food to the table of better-off middle- and working-class citizens is thus a foundational issue for food ethics. At one level, the injustice of this situation is patently obvious to anyone. Like the viewers of Murrow's documentary, the readers of Esterbrook's *Tomatoland* or Patel's *Stuffed and Starved* cannot fail to grasp the core moral significance of food justice. Yet few (if any) efforts to bring these issues to the attention of the public have been as successful in reaching a large audience as *Harvest of Shame*. Consciousness-raising exposés will continue to be a prerequisite for any sustained effort to analyze and discuss the ethical issues that arise in connection with food production, distribution, and consumption. Such work is best achieved by investigative journalists who concentrate on the facts yet report them in a manner that stimulates moral outrage and motivates action. Fortunately, there are many authors now engaged in efforts to raise the consciousness of a complacent public. In addition to Esterbrook and Patel, we might note Robert Gottleib and Anupama Joshi's *Food Justice*, Oren Hesterman's *Fair Food*, or *Hope's Edge* by Frances Moore Lappé and Anna Blythe Lappé. All three of these books are supported by social media efforts and organized campaigns to keep readers informed and mobilize action.

This book is not intended to duplicate or compete with the important work being done by this cadre of writers and social activists (and there are many more). It is possible to do a great deal of good work in response to these issues without the benefit of much analysis or reflection on the nature of ethics, the theory of justice, or the philosophy of food. Unfortunately, it is also possible to do harm, or at least to expend energy on efforts that are unlikely to have enduring impact. Activism in response to obvious injustice rapidly encounters lassitude and prejudice that retards the progress of reform. *Harvest of Shame* was deliberate in its inclusion of segments on white migrant workers, as well as black and Hispanic workers. Murrow and his collaborators at CBS News were mimicking a strategy for enlisting the sympathy of white viewers that was deployed even earlier by John Steinbeck. Steinbeck followed migrant laborers in the camps of depression-era California for a series of articles for the *San Francisco News* in 1936. His reporting was later novelized in *The Grapes of Wrath*. Steinbeck had hoped that a focus on white migrants would help his readers get beyond the racist attitudes that, he assumed, were making them unresponsive to the cruelty routinely experienced by black, Hispanic, and Asian minorities.[5] It may be in the steps taken *beyond* the initial emotional outrage following the exposure of injustice that philosophy can start to help.

However, given this history of journalism and activism in response to injustices in the food system, one *might* question whether there is really a philosophical problem at all. On the one hand, the exposure of these atrocious practices is seldom met with any defense. It is as if people possess a native ability to recognize injustice when they see it. They do not really need a philosopher to point it out to them. There is thus a sense in which offering a philosophical analysis of injustice in the food system is simply pointless. Everybody already knows, and nothing that a philosopher can say will do anything to further illuminate the problem. Perhaps philosophers or moral psychologists could say something about what motivates people to do something about an injustice once they see it, but it would seem unlikely that there would be anything special about injustice in the food system. The relevant body of theory would have to do with motivation in general. It is not as if people are especially lacking in motivation when it comes to unfairness or abusive treatment on farms, in slaughterhouses, in food distribution, or in retail grocery stores and restaurants. Given this line of thinking, it is not at all clear that the injustices in the food system are *philosophical* problems, and thus it is unlikely that they would be improved or resolved by a philosopher focused on food ethics.

At the same time, the things that a philosopher could say about these problems will very likely reflect theories of social justice that have been developed with little thought to the particular problems that arise in the food system. One might, for example, argue that both income and access to food are what John Rawls called primary goods. A primary good is one that will be instrumental toward realizing any and all of the diverse plans of life that individuals in well-ordered societies pursue. It thus follows from Rawls's theory of justice that inequality in the distribution of primary goods can be justified only when that inequality benefits the worst-off group.[6] There may be a little philosophical work to do in showing how current food-system practices fail to be supported by this rule (which Rawls called "the difference principle"), but all the conceptual heavy lifting is being done by Rawls's version of a neo-Kantian rights theory. There's really nothing more interesting or philosophically significant about applying it to food-system workers than to the homeless, the perennially unemployed, or to other low-wage workers in sectors such as custodial work. All need some recognition of their right to primary goods.

Alternatively, one can take a libertarian view of justice. Applications of a libertarian view might start by noting that we generally reserve the word *justice* for social and political problems that require some response from government. The libertarian perspective has tended to support minimalist interpretations of government's role in achieving ethical ends. Following Mill's argument in *On Liberty*, the state has a justified function in protecting people whose basic liberty is infringed by the acts of others. Farm workers being subjected to a form of slavery would obviously meet that test, but a libertarian will very likely conclude that many of the other questions here turn on whether individuals have a duty of charity with respect to food-system workers or other needy people. Perhaps private citizens *should* be voluntarily offering gratuities to food-system workers who are badly paid or contributing to the local food bank in order to help hungry families. But these are acts of personal charity, not matters of justice requiring action by government. The libertarian approach to justice will be more focused on the ethical question of whether state action and public funds can permissibly be used to address the moral problems chronicled above. In some of the more extreme cases—tomato workers imprisoned against their will—the answer will certainly be "yes." This kind of abuse is a proper target of police action, and offenders should be pursued through the courts. But libertarians might view initiatives to raise the wages of food-system workers a bit more critically. This kind of action uses state power to implement the redistribution of wealth from one person's pocket into another's. That will violate a basic principle of justice as libertarians see it. Of course these comments on libertarian justice are intended to be illustrative rather than definitive. The larger point goes beyond the way that libertarian theorists work out the application of their principles to a food-system case. Here things are similar to Rawls's more egalitarian view: once again there is nothing particularly special about the fact that we are analyzing the food system. We would be making the same judgments about being coerced and abused in textile manufacturing or the sex trade, and if we are consistent libertarians, we will say the same things about the way that public funds are used in these other areas of the economy.

And it is not clear that this general point changes if one shifts to more radical social theories—to Marxism or any of its close relatives. To be sure, a student of Marx will be sorely exercised by the injustices that have been outlined above, and a socialist of any kind will very like support state programs that a libertarian would reject. But when asked whether there is anything particularly distinctive about the fact that these abuses are occurring in the production and distribution of food, it is difficult to imagine how a Marxist would find them to differ from inequalities in access to employment, health care, or education. For a theorist to the left of Rawls, the problems are very likely symptomatic of capitalism in general. The theory has already explained why the logic of capital accumulation determines how the owners of these tomato fields, processing plants, grocery stores, and restaurants will treat their workers.

Marx himself explained why those at the bottom rung of the wage scale cannot expect to earn enough to feed their families when wage rates are being set by market forces. There is nothing in that explanation that singles out food-related firms (or indeed firms of any particular kind). So the general point to take from all this is simply that while exploitation and injustice may well be a problem in the food sector, the mainstream philosophical approaches seem to be fully capable of accounting for when, why, and whether there is injustice without having to say anything about the fact that it is food we are talking about. It is in this sense that someone might say that there is nothing *philosophically* interesting about social justice in the domain of food ethics.

Race, Gender, and Ethnicity

But perhaps there is more. Although poor whites do suffer from exploitation in today's food industry, a more adequate analysis of injustice in the contemporary food system recognizes that a legacy of overt racism and ethnic prejudice played a key role in making the global food system what it is today. This history has roots in European colonialism. The quest for colonial empire may have begun in pursuit of gold and natural resources, but it was fueled by a search for cheap sources of food for workers in the burgeoning factories of industrial England and France.[7] The first step, of course, was taking the land itself from whatever indigenous people happened to find themselves in the European path of conquest. The second step was making that land work for the benefit of Europeans. The enslavement of African blacks throughout the Americas was largely for providing labor on plantations growing food and fiber crops. In the aftermath of the American Civil War, the system of slave labor was replaced with sharecropping, which kept blacks in perpetual poverty through a repressive system of "furnish" (or credit for supplies) coupled with often dishonest compensation for blacks' share of the annual crop. Meanwhile, Western ranching and fruit and vegetable production grew around the use recent immigrants and undocumented laborers, initially from Asia but later primarily from Mexico and Central America.[8] There is also a gender dimension to the exploitation in the current food system. Women are disproportionately employed in low-wage food industry jobs, and their role as a decision maker on American farms has been persistently denigrated, ignored, and even repressed by extension services, farm service firms, and banks.[9] Internationally, women farmers outnumber men several times over, yet women seldom have equal access to farm credit and may lack equal legal protection for land tenure.[10]

In addition to these gender-based economic and social injustices and inequalities in the food industry, there are injustices to be noted among marginalized groups' access to food. This is, of course, most evident with respect to global food security, a topic that this book revisits several times. However, even when poor people or racial and ethnic minorities obtain an adequate supply of dietary calories, they may lack access to fresh fruits and vegetables and or to foods that are culturally appropriate. They may find themselves in urban "food deserts," or relatively large urban areas without grocery stores. Available food may come from fast-food restaurants and convenience stores and may be highly processed food high in fats, salt, and sugar. First lady Michelle Obama has argued that this pattern of urban economic development has longstanding consequences for children's health. She has launched the "Let's Move" program to promote better diet and more exercise, but she has also castigated the food industry for failing to provide children in poor urban neighborhoods with healthier alternatives.

There are thus reasons to suspect that mainstream ideas about social justice may not be fully up to the task of analyzing food justice. The last four decades of work in social theory and philosophy have seen the emergence of feminism and critical race theory, both of which have challenged white male privilege in the traditions of social justice that continue to dominate in Western culture. One important theme in this work concerns environmental justice. Work in the 1980s by sociologists such as Robert Bullard, demographers such as Bunyan Bryant, and the United Church of Christ's Commission for Racial Justice documented the way that racial

minorities were bearing a disproportionate share of the toxic exposure and health burden from environmental pollution. Although
the idea of "environmental racism" became the subject of debate
in the 1990s, philosophers such as Robert Figueroa and Kristen
Shrader-Frechette have argued that irrespective of whether a racial
motive can be proven, there are important issues of justice to be
recognized with respect to environmental issues. The privation,
displacement, victimization, and isolation that befall people in a
pattern that tracks racial or gendered identities raise issues of justice in at least four fundamental ways.

First, *distributive environmental justice* refers to the disproportionate exposure to environmental injury and insult by the poor
and by groups who have historically been victims of other, more
conventional types of social inequality. Since economic inequality, gender, and ethnic identity are not morally relevant factors
with respect to health or environmental risk (as voluntary behaviors such as smoking might be), statistically significant burden of
disease and low-quality living environments endured by people in
these groups is unjust. Second, *participatory environmental justice*
refers to the opportunity for people in these groups to have an
effective voice in decision making about the factors that contribute
to environmental risks and to environmental quality of life. When
people in marginalized groups are systematically excluded from
decision making (perhaps in virtue of the way that they are prevented from achieving political power or higher educational expertise), there is an injustice. *Justice in recognition* aims to acknowledge
the identity and culture of marginalized people rather than sup-
posing that the generic characteristics of the "rational man" that
motivated classical European theories of justice is a one-size-fits-all
universal. Finally, *restorative environmental justice* concerns efforts
that respond to the historical and current burden borne by the victims of past injustice and seek compensation.[11]

There are reasons to think that work on environmental justice
may be applicable to some of the aforementioned issues in food justice. The phenomenon of food deserts is especially amenable to an
analysis in terms of distributive environmental justice, and may

also reflect minority groups' lack of voice in food industry decisions about where to locate grocery stores, farmers' markets, and other outlets for quality food. The core issue may be one of participatory environmental justice, as civil authorities and major food-system firms make fatal choices without hearing from the marginalized groups most seriously affected. What is more, many analyses of food security presume that food needs can be fully characterized in physiological terms. They stress individuals' need for calories and specific nutrients, for example. Yet this may be a profound misrecognition of who they experience themselves to be. Cultural identities are constructed through the repeated consumption of certain foods or recipes, and through dietary or food preparation practices. An approach to food justice that focuses on meeting biological needs may well perpetuate acts of oppression that separated racial or ethnic groups from their cultural traditions in the first place. Such failure to recognize a person's or group's identity is an injustice in itself that is compounded by inappropriate restorative measures. In this way, even well motivated efforts to address the food security needs of marginalized groups simply repeat the original injustice of misrecognition.

But environmental justice is not the only concept that might be applicable to the problems of social inequality or racism with respect to food. In fact, a plethora of new concepts and theoretical orientations have emerged for better characterizing and theorizing the ethics of social issues—wicked problems, political ecology, postcolonialism, intersectionality, phalogocentrism, ecofeminsim—the list is indefinitely long. Other philosophical traditions might frame issues in terms of obligations to future generations, discursive dilemmas, collective intentions, or the non-identity problem. There is a lot of serious work to be done in reimagining the domain of social justice and achieving adequate awareness of food justice at the theoretical level. Although I hope that the ideas developed throughout this book will be complementary, giving each of these new ways of thinking and writing their due is impossible. We must come back to a point made in considering the applicability of mainstream philosophical thinking on social justice: none of these recent innovations have taken food as a central topic. Even if we can *extend* them to food ethics, many of the most promising ideas in the new literature are at best tangentially related to food. As before, many of the social issues noted above might also be (and have been) raised in connection with health care and social services access, labor relations in textile production and retailing, or low-skill jobs in the service economy. Issues of racial and gender inequality are also occurring in the manufacturing sector as globalization negates important gains made by the labor movement during the first half of the twentieth century. Food production, distribution, and consumption is a social domain in which many pervasive social issues involving race, ethnicity, and gender are especially evident, but the food industry may not be uniquely significant. What is more, neither the critics of older views in the tradition of liberalism and utilitarian ethics nor the theorists offering new perspectives have chosen to develop their work in connection to food. Thus, though important work linking these traditions to food cries out to be done, it may be more crucial at this juncture to shift our attention to topics that *are* more demonstrably distinctive for growing, processing, and eating our food.

The Food Movement

If issues of injustice are pervasive and longstanding, one might ask why the recent spate of journalism and social activism in connection with food consumption has garnered as much attention as it has. Social trends are to some extent the result of happenstance: what seems to be a pattern may simply be the coincidental occurrence of otherwise unconnected events. In contrast to short-lived social movements, the struggle over social justice is, in an important sense, the storyline for all of human history. Well before the beginnings of recorded history, bands of human beings formed cultural norms that helped them cooperate and survive as a group. These norms were handed down from one generation to another and became the backbone of a transmissible social identity.

But within-group and inter-group striving for a better life also caused violent conflict in the form of warfare, oppression, theft, and slavery. For at least three thousand years, human cultures have developed increasingly sophisticated institutions for ameliorating conflict through negotiation, debate, and social learning. In the eighteenth and nineteenth centuries, people began to notice a pattern of evolutionary growth in these institutions that became the basis for an ideal of social progress.

But while thought leaders of the Enlightenment era were converging on the idea of progressive social change, a new kind of competition and conflict was emerging between people who were committed to one set of social institutions rather than another, and to the corresponding ideals of progress that were associated with each. In one sense, world history since 1800 continues the story of competition for resources and the quest for domination that gave rise to violent conflict among peoples in prehistory. Yet there is also a sense in which our recent history is more adequately written as a struggle of ideas or as a philosophical debate over the way to organize our societies. One theme in this story is the way that a constrained kind of competition in the marketplace can unleash forms of innovation that achieve hitherto unimagined forms of human betterment. Advocates of this ideal see a very specific role for government in enforcing "the rule of law": preserving basic liberties, adjudicating disputes, and protecting the rights of property owners. But others point out that if enforcing "the rule of law" results in systemic poverty and deprivation for some groups, this cannot be an adequate conception of social progress. This is especially so if these groups begin the marketplace competition from a significantly disadvantaged position. The injustice is morally worse when these disadvantages are the result of earlier or unpunished violations of the very principles advocates of limited government call upon to support their position.

Simply put, there is a longstanding disagreement over what social justice requires, and what role government should take in achieving it. This controversy will shift from one focus to another depending on a complex and largely unpredictable mix of factors.

From the late eighteenth century well into the nineteenth century, one key focal point has been the way that class identity has excluded some groups from equal protection under the law. It is now difficult for many to recall that this debate once centered on the idea of a nobility—the dukes, earls, and barons who controlled the administration of law in most nations. The dominance of a noble class was initially challenged when other white, male landowners were admitted to the inner circle. The battle for social justice has successively attacked barriers to full participation and to equal protection under the law that have been defined by wealth, gender, and racial identity. At the same time, the battles have often revolved around specific sectors of the economy. For a century or more, the focus was on labor, first with respect to the abolition of slavery and later with respect to the rights of factory workers to organize and bargain collectively. Advocates for social justice have always seized the moment when something captures the public's attention: the Pullman Strike in 1894, the 1907 explosion at the Fairmont Coal Company, the Triangle Shirtwaist Factory fire in 1911. A shift in emphasis from heavy industry to mining or textile workers occurs when publicity creates the opportunity for progress in a particular domain.

At some junctures, the public's attention has been gained because a talented writer has entered the scene. Early reforms in the food industry were a response to Upton Sinclair's exposé of Chicago meat-packers in his novel *The Jungle*. Sinclair intended his book as a contribution to worker's rights and social justice. He described workplace accidents in which people who fell into rendering tanks were ground into the product. The primary impact of the book was the 1906 passage of the Pure Food and Drug Act, which famously led Sinclair to remark that he had aimed for America's heart but hit its stomach. Steinbeck's account of the Joad family and Murrow's high-profile television journalism also had the effect of turning social justice advocates' attention to the food system for a time. There is little doubt that two highly effective books in the early years of the twenty-first century have played a significant role in bringing food to a wider public notice. *Fast Food Nation* (2001) by Eric Schlosser summarized the history of developments in the food industry since World War II and juxtaposed discussions of the working conditions for beef packers and the emergence of synthetic flavor chemistry.[12] Schlosser painted a picture of a profit-driven industry with little ethical concern for its workers or its customers.

Michael Pollan's *The Omnivore's Dilemma* was published three years later. Pollan structured his book as a philosophical inquiry into the ethics of eating, casting himself as a naïf who wanders through the American food system, discovering its complexities and the dilemmas faced by anyone who hopes to eat in good conscience along the way. Social justice does not figure prominently in Pollan's discussion of the industrial food system, but like Schlosser, he shows that the current configuration of technology and business practice has the effect of concealing from the consumer much of what happens throughout the food supply chain.[13] There is much to admire and even more to ponder in Pollan's work, but it is important to realize that *The Omnivore's Dilemma* adopts the posture of ethics as a rhetorical device. Pollan is a journalist who has reconstructed his account of the food system after many hours of reading and interviewing farmers, businessmen, and academic experts. He builds his readers' interest by developing a personal journey narrative in which he is the main character. The narrator is a likable chap who slowly and reluctantly uncovers what Schlosser proclaims on the front cover in his subtitle: the dark side of the American meal. Julie Guthman, a professor at the University of California Santa Cruz, has faulted Pollan for covering up the way that he is, in fact, relying on what others have told him rather than simply discovering things for himself, and that in doing so he not only fails to give adequate credit to his sources, he also shifts the emphasis so that social justice hardly appears at all.[14]

The story that I have been telling here situates the work of journalists such as Schlosser and Pollan within a larger arc of struggle and debate over social progress. It would also be possible to tell the story of the food movement as a relatively minor shift in popular culture. For example, we could start by pointing out that as incomes in industrial societies start to rise, food becomes a form of entertainment. People eat meals out of the home for fun, and they try new recipes as a way to put a bit more variety into their daily routine. It would be hard to ignore that surge of magazines and television programs dedicated to cooking or exploring fine and exotic dining experiences. While it is possible to interpret the growth in gardening, farmers' markets, and artisanal restaurants as a form of cultural renewal or resistance to the powerful food industry, it may be more plausible to view it as the latest fad: wealthy people trying to amuse themselves by playing peasant games. Certainly some critics of the food movement see it as an elitist phenomenon having absolutely *nothing* to do with social justice. Guthman's criticism of Pollan is that he makes it easy for this kind of interpretation to arise.

My own view is that the "elitist fad" interpretation and the "social movement" interpretation are not incompatible. The faddists are certainly not likely to have staying power for the causes that activists are pursuing, but their mere presence can give a temporary impetus to the prospects for progressive change. Sometimes a temporary impetus is all that is needed for change to occur. What is more interesting is to examine how ethics of production and consumption in the food sector diverge from the more traditional analyses of social justice that might be given with respect to labor, health care, or any number of other potential areas where social class, ethnicity, race, and gender are linked to systemic oppression, deprivation, and exploitation. Food resonates on many levels. The food that we eat passes through us and to some extent becomes incorporated into our very bodies. At the same time, food links us to family and cultural traditions. It is not surprising that we react with disgust and distaste upon learning that food we have eaten is not what we thought it was. These emotional reactions are available to those who are advocates for a progressive vision of social justice. To the extent that self-interest or disgust can be linked with social justice, social action can be mobilized in a way that would otherwise be difficult to achieve. But of course the people who are just in it for a good time will eventually tire of playing at food justice and turn to something new for their amusement. And there is always the risk that, as Upton Sinclair learned, a critique mounted in order to promote social justice will be deflected when questions about the safety, purity, or authenticity of food come to the fore.

On the one hand, food has a powerful *integrative* effect on the moral imagination. Otherwise diverse and distinct social problems come together around food. Securing something to eat is a biological necessity. It would be surprising if the cognitive architecture of the brain had not evolved so that emotions both positively and negatively reinforce a suite of eating behaviors in a manner that increases the survival chances for both individuals and groups. The emotional underpinning of food behavior may encourage mental leaps and even lapses of logic when food norms are threatened or questioned. Social psychologists have argued that food is a domain where "magical thinking" predominates. Metaphor becomes metonymy, and cognitive categories that may have their basis in social habits appear to take on deep metaphysical significance.[15] The upshot is that distinct categories of harm become blurred. Problems in the treatment of workers become threats to personal health. Cultural faux pas become environmental threats. Blurred categories and mental leaps have contributed to the recent emergence of a bona-fide food movement in recent years, dedicated ambiguously to an assault on the causes of obesity; harm to animals; environmental decay; and injustice to women, minorities, and other marginalized groups.[16] It is not immediately obvious what we should think about this from the standpoint of ethics. Philosophers have traditionally insisted upon strict logical rigor—keeping the peas from touching the mashed potatoes, to repeat my food metaphor—but perhaps an evolutionary account of the mind would show us that these heuristics and biases confer adaptive advantages that have helped human communities survive. The food movement is progressing as if the political efficacy that is associated with hazy thinking and the integrative power of food confers legitimacy on actions that resist oppressive acts in the food system.

On the other hand, the diffuseness of these concerns and the Balkanization of both markets and policy suggests a problem for action on any of the many issues. Policies intended to counter-act rising rates of diabetes may have little impact on the social or political marginalization of women and minorities, and vice versa. Indeed, there is a real possibility that action to redress one type of ethical problem will exacerbate problems in another domain. Thus when Upton Sinclair, who ran for the governorship in California under the EPIC campaign banner in 1934,[17] complained that he had hit America's stomach instead of its heart, he should be seen as offering a warning to today's food movement. The wheels may well come off the food movement when ethical vegetarians and environmental advocates find themselves on opposite sides of an issue. History teaches that the Pure Food and Drug Act of 1906 was a landmark piece of legislation that created the global model for government action to ensure a safe and wholesome food supply, but it also gave rise to a highly reductionist interpretation of food safety risk that is wholly decoupled from environmental or nutri-tional hazards, not to mention social impact and the sociocultural meanings that may form the core of our emotional attachments to food. One could argue that its very success was achieved at the expense of a more integrated and holistic view that might have allowed progress along a broader front of social issues.

Food Security and Food Sovereignty

The theme of integration brings us to one last topic under the head-ing of social injustice. As Chapter 4 discusses at more length, there is widespread recognition that food is, in one sense at least, *quite* different from many other social goods. Everyone must have food to survive, and when inequality in access to food reaches an extreme point, people die of starvation. Well before the point of starvation, children are especially vulnerable to the diseases of hunger and food deprivation. The Food and Agricultural Organization (FAO) of the United Nations maintains a "hunger portal" that reports global statistics on undernourishment and encourages visitors to delve more deeply into the complex causes of hunger on a global basis. As of this writing, FAO notes areas of progress on the global front but still reports approximately 842 million people in the world subsisting with less than adequate food.[18] There is thus a stark sense in which secure access to food is morally compelling in a manner that transcends philosophical argument. This fact is acknowledged in the United Nations International Declaration of Human Rights. As noted already, the policy-oriented literature that has grown up around the imperative of securing universal access to food has defined the goal as "food security."

It is not possible to do justice to the way that food security has been approached by economists, nutritionists, and food policy specialists over the last fifty years. It must suffice to say that the core notion has been a person's ability to consume a diet that is nutritionally adequate, given current age, gender, daily expenditure of calories, and general state of health. Malnutrition can surface in the form of inadequate calories or an imbalance of nutrients, the latter being the case when diets become insufficiently varied. Anyone at a significant risk of being unable to consume a nutritionally adequate diet over a specified time horizon would be considered "food insecure." But food insecurity can take many forms. Victims of natural disasters such as a typhoon, drought, volcanic eruption, or tsunami experience periods of extreme but short-lived food insecurity. Their needs can often be met by emergency shipments of food through the World Food Program, one of FAOs more capable agencies. Ironically, such victims of disaster stimulate the most compelling moral response on the part of the food secure. To be sure, there is a genuine need. Nevertheless, although it may take a year or more for local agriculture to recover, these are the most solvable problems in food security. The more intractable ethical problems of food security concern people whose poverty prevents them from having access to adequate diets even when food is available and plentiful in local markets.[19]

However, there is also a line of thought that runs like this: a *nation* will not be secure if it is vulnerable to short-term collapses in the food supply. The tradition of political economy has tended to evaluate food security in terms of nation's ability to produce enough to supply their military garrisons and to have secure supply lines to them. This way of thinking about the ethical/political significance of food systems goes back to the ancient world, but it was forcefully articulated by James Harrington, an English theorist of social justice in the seventeenth century. Harrington stressed the connection between military security and a nation's ability to feed its army, linking both to more traditional arguments praising the moral character of farmers.[20] Such arguments surface again in Chapter 6. Here notice that it is not so much a natural disaster as the potential for war that drives Harrington's thinking on food security. There is a longstanding policy tradition that has been skeptical of trade as a reliable source of food and thus a guarantor of food security. Whether or not this skepticism is justifiable, it has carried over into present-day debates on food security in the form a dual conception. On the one hand, *people* enjoy food security when they have access to a nutritionally adequate diet; on the other hand, a *nation* is food secure when it has secure access to enough food for everyone that it wants to feed. And today as in the classical period of political economy, there is skepticism about the reliability of international trade.

Over the last decade, a new voice has entered this debate. Via Campesina began as a loose-knit Latin American group representing a variety of peasant farming interests. It has expanded and now reflects the viewpoint of poor or marginalized people across the globe. Farmers do have a special kind of vulnerability when food security is being negotiated. If policymakers in the national capital *do* decide that food security (in either sense) can be guaranteed by trading on international markets, local farmers may be facing stiff competition from imported grain—grain whose production may have been subsidized in rich countries. In the meantime, policymakers in the national capital may be hoping that local farmers will stop growing subsistence crops and switch to something like cocoa, coffee, or biofuels—crops that generate foreign exchange. Advocates for Via Campesina started to resist the idea of food security. *Their* interest is not simply one of having access to a nutritionally adequate diet. Indeed, what they fear most is being placed in a position where they must earn money to purchase a nutritionally adequate diet grown on foreign soil. So Via Campesina has argued that what matters morally is *food sovereignty*—the ability to control the structure and organization of one's local food system.[21]

Unfortunately, as compelling as this argument is for the small farmers that Via Campesina represents, it is not an idea that necessarily travels well. Some have presumed that food sovereignty is just another label for the long-lived way of defining food security at a national level, and they have then taken the view that this cannot be achieved unless a nation produces enough to feed its entire population within its national boundaries. No European nations are food secure by this standard, and none are taking steps to achieve it. Thus, there is some degree of skepticism about food sovereignty among the policy specialists that have focused on getting nutritionally adequate diets to poor people for the last fifty years. Others have seized the term as a banner for all manner of social justice activism in the food systems of the industrial world. Food sovereignty then becomes a moniker for increasing the wages of fast-food workers and for labor organizers in the fields with tomato pickers. In some quarters, food sovereignty is advanced as a reason why traditional farmers and artisanal food processors ought to be exempt from national food safety regulations. As ethically laudable as these causes may be, it is far from clear how they meld with the goals of Latin American peasant farmers who are trying to preserve both their local markets and a farming way of life.

Nevertheless, there does seem to be something philosophically important now being advanced under the flag of food sovereignty. For Via Campesina, at least, the idea points to the way entire rural communities, local cultures, and longstanding social relations are brought together through the production, preparation, and consumption of food. If you are living in a Latin American village where most people are farmers, it is easy to see how the production, preparation, and consumption of food integrates families into a community. The continuance of the community depends upon people caring for one another and looking after the ways that they have long endured through the production, preparation, and consumption of their food. The survival and maintenance of these food practices is critical to the sustainability of these communities in every sense of the word. It is the basis of their economic livelihood. It is what binds them to their natural environment, and one can be sure that they have a keen sense of what is needed to preserve it. But finally, it is in the performance of these practices that the institutions and relationships of the local community are reproduced over time. Destruction of this nexus of institutions and material practices means destruction of the community. Individuals might survive, to be sure. They might even move to the city and work in a factory where their net income will be higher than it is today. They might achieve what FAO calls food security by doing this, but would that be justice?

A Temporary Conclusion

The facts clearly show that the industrial food system relies heavily on low-wage labor, and that food industry firms—including farms, processors, and retailers—are well represented among the bad actors of labor relations. Bad-faith negotiations, exploitation and violation of work rules, and even felony violations of basic civil and human rights are all too common in the industrial food system. No theory of justice would countenance the more extreme acts of injustice perpetrated by food industry firms, but they are able to get away with it largely because the people who are victimized by these practices often find themselves in desperate straits due to poverty, lack of education, and vulnerabilities that relate to their immigration status, work history, or family situations. From the perspective of these workers, there is no sharp distinction between being victimized by violations of the law and legally sanctioned victimization through part-time, minimum-wage employment and work rules that keep people at or near the poverty level.

While Rawlsian or Marxist conceptions of justice would find the treatment of food-system workers in many countries (and especially the United States) to be inconsistent with the philosophical dictates of social justice, there are other views. Farming and food-system employment have always been poorly compensated, in part because the skills needed for this work have historically been virtually ubiquitous. When anyone could do the manual work of farming, food processing and cooking, there was no reason for employers to pay more than the lowest wage that the market would bear. And even as economic growth and social change have produced a working class having little to no familiarity with farming and food production, there have always been plenty of immigrants with farm backgrounds who are eager to take on these jobs. If this is just the workings of the labor market, then philosophers with a certain kind of libertarian or neoliberal mindset may find the circumstances of food-system workers to be regrettable but still not a true injustice. The appropriate response will occur when enough of them attain the education that will allow them to compete for better jobs. When no one will take these food-system jobs because there are better jobs to be had, scarcity in the labor pool will force food industry firms to pay more.

There *is* work to be done in discerning how these contrasting philosophical approaches to economic justice would apply to the industrial food system, but the picture that I have just drawn suggests that the philosophical problem of justice is one that must be addressed at a level of abstraction where it is largely irrelevant that we are talking about food-system employment, as distinct from employment in building construction, manufacturing, or some other sector of the industrial economy. Indeed, the economic situation of fast-food workers or grocery employees would seem to have everything to do with the structure of retail employment and very little to do with the fact that these businesses happen to be engaged in selling food products to retail customers. So, without denying the social relevance or immediacy of these issues, we can conclude that these problems of social justice would not give rise to a form of food ethics that would be different from every other application of the theory of justice.

It is less clear that traditional theories of justice articulate what is at stake for the peasant farming villagers of Via Campesina. On the one hand, contemporary social theorists who share Marx's skepticism of capitalism have argued that the global food system should be understood as a systemic nexus that functions to extract the last shred of economic value from the production, distribution, and consumption of food and to repress any form of political action against it. Everyone needs to eat. Therefore, control of food is a powerful locus for both profit-taking and the exercise of social control. Food sovereignty is, on this view, a mode of resistance against this totalizing food regime. "Food politics" starts to look like something that is both novel and deeply significant for the way that it cuts to the heart of everything that matters. On the other hand, the driving forces that are identified by this analysis may not be very new at all: neoliberal promotion of trade and global institutions such as the World Bank are pushing toward a kind of food security that threatens small farming villages, and the big winners are corporations who first manufacture the seeds and chemicals and then control the trade in grain and manufactured food products right down to the retail level. Opponents of global capital and advocates of small farmers have a ready-made common enemy. Once again, the argument starts to circle back to a common theme. It is starting to look like the food connection is rather accidental. A Marxist is not surprised to find issues of justice in a neoliberal corporate food regime, but they are symptomatic of issues that exist *throughout* a capitalist economy. Food winds up merely as a convenient tool for raising peoples' consciousness of a more widespread type of oppression.

There is, in conclusion, a sociological question: is "the new politics of food" a *transformational* social movement, or will it exhaust itself in a few political reforms that leave the overall structure of contemporary society intact?[22] If the latter, the food movement will be like the labor movement, the civil rights movement, and the women's movement before it. In either case, there are philosophical questions that demand an answer. Those who look for transformational power in the new emphasis on food may see these issues as having a salience or universality that earlier social movements have lacked. If the food movement achieves significant but nontransformative change, it will still be important to articulate why food practices involve matters of ethical importance. The chapters that follow offer suggestions for articulating and analyzing several key topics in food ethics, but leave open the question of whether food sovereignty amounts to a comprehensive or transformational approach to problems in social justice.

In a nutshell, it may be the way that food moves *across* so many distinct and distinguishable contested terrains that provides a germ of insight. The intersectional nature of food-system issues is the key lesson to take from a review of social justice activism and the continuing injustices in farming and food industry firms. It is the way that food brings these issues together that matters for social movements, that gives them salience and some hope for effectiveness. In writing "The Food Movement Rising," Michael Pollan may have offered his answer to critics like Julie Guthman. He first quotes sociologist Troy Duster:

> "No movement is as coherent and integrated as it seems from afar," he says, "and no movement is as incoherent and fractured as it seems from up close." Viewed from a middle distance, then, the food movement coalesces around the recognition that today's food and farming economy is "unsustainable"— that it can't go on in its current form much longer without courting a breakdown of some kind, whether environmental, economic, or both.[23]

Philosophical ethics must find a way to shake off its reductionist tendencies to see that. It must undertake a wide-ranging overview of diverse food-system topics to appreciate the connections. Perhaps then it will be possible to link sustainability and social injustice. If I am right, the seemingly unrelated series of topics in the following chapters may be more closely connected than they appear.

Chapter 3

The Ethics of Diet and Obesity

Many years ago, I was seated next to an in-law at a family gathering. Upon learning that I was a philosophy graduate student working on ethics and food, she immediately wondered whether I could tell her anything that would help her lose weight. I couldn't then and I still can't now, and I was stupefied by her inference that philosophers can be expected to know something about weight loss. What I learned from this encounter is the *personal* nature of food ethics. People want a food ethics to tell them what they should eat. In Chapter 1 we undertook an initial exploration of the way that food purchases affect others, but my dinner partner's question may have been more to the point. How does what I eat affect *me*? My companion may have been interested in her appearance—at least that's what I thought—but many of us are now concerned with how what we eat is linked to chronic debilitating diseases. Diabetes and heart disease are among the leading causes of death in affluent societies, and overweight people have higher than average chances of suffering from both. If we are going to ask better questions about our diets, it is useful to examine how the ethics of diet has undergone a number of surprising twists and turns over the millennia, especially in recent decades.

Philosophical Dietetics: The Ancient and Medieval World

Philosophers of the ancient and medieval world were attentive to diet. The moral legitimacy of consuming animal flesh was certainly one topic of enduring interest,[1] but healthful dietetic practice might well have been viewed as a natural component of ethics in the ancient world. In fact, promoting health might not have been seen as all that different from respecting animal life in the world-view of ancient Greece and Rome. Both Stoic and Epicurean schools of philosophy devoted significant attention to dietary practice. They viewed the cultivation and control of appetite as a spiritual practice and as a locus of virtue or vice. Present-day talk about food continues to reflect ancient doctrines, though sometimes in ways that diverge dramatically from the actual teachings of the sages. For example, in contemporary speech, an *epicure* is someone who is able to appreciate the sensory qualities of a well-prepared meal. The word derives from Epicurus (341–270 BCE), a phi-losopher who rose to prominence in Athens a generation after Aristotle. As the Epicurean school of philosophy has come down to us, pleasure is the chief aim of life. This squares up with the worldview of many present-day epicures, who can wax poetic at the gustatory delights of good food, gourmet cooking, and the pre-sentation or ambiance of a fine dining experience. For Epicurus himself, however, the road to pleasure lay not in indulgence but in the cultivation of *ataraxia*—a virtue that might be character-ized as self-reliance, save for the fact that many people today would associate self-reliance with excessive wealth and individual achievement. Epicurean pleasure was achieved through a living a life in which one was free from pain and fear, to be sure, but the route to this life charted by Epicurus was to achieve satisfaction by matching one's consumption to one's means and by enjoying the company of friends. There is little evidence that consumption of food was a particular focus of ataraxia, though Epicurus appar-ently endorsed intervals of frugality as a way to prepare oneself for the possibility of hard times.[2]

The contemporary meaning of *stoic* indicates an ability to endure hardship and pain without showing emotion. Stoic philos-ophers of the ancient world did endorse both dietary and sexual restraint as a form of *ascesis*, associated with our word *asceti-cism*. Ascetic practices were intended to cultivate an ability to discipline one's appetites in general, and were understood within the context of achieving the virtue of *sophrosyne*. *Sophrosyne* is a key idea that runs throughout virtually all schools of ancient Greek philosophy. It is often translated as "temperance," though this, too, will be misleading if temperance is too closely associated with various temperance movements of the twentieth century, for whom it meant total abstinence from alcohol. *Sophrosyne* implies balance—a life or personality that is resistant to vices of excess. And for the Greeks, *most* vices represented unchecked tendencies that had been allowed to run amok. In the case of diet, we might think of *sophrosyne* as a balance between abstemious or "picky" eating and gluttony. A dietary *ascesis* could be a practice of reminding oneself of the need for balance, and might well take the form of staying just a little bit hungry. Writings of the Roman Stoic philosopher Seneca (4 BCE–65 CE) do contain references to food to illustrate how a wise person achieves an ability to discipline the sway of passion. Dietetics was thus a component of Stoic philosophy, and *ascesis* was a practice designed to discipline appetite, understood both literally as the urge to eat but also more generally as the tendency to be carried away by desires of all kinds.

Dietetic injunctions also surface as moral themes in the medieval period. Gluttony appears as one of seven deadly sins. Pope Gregory I is said to have characterized gluttony as "an inordinate and selfish desire for the sensual pleasure which arises from the gratification of taking food, when such desire develops into action."[3] The five ways to commit gluttony were summarized in the Latin aphorism "*Laute, Nimis, Studiose, Praepropere, Ardenter.*" St. Thomas Aquinas summarized the five types as follows:

> *Laute*—eating food that is too luxurious, exotic, or costly
> *Nimis*—eating food that is excessive in quantity
> *Studiose*—eating food that is too daintily or elaborately
>   prepared
> *Praepropere*—eating too soon, or at an inappropriate time
> *Ardenter*—eating too eagerly[4]

Some scholars have argued that interpretations of gluttony within the context of the seven deadly sins miss the fact that during the Middle Ages gluttony was understood to be more closely associated with drunkenness than overeating. On this view, medieval remonstrations against gluttony might be more accurately associated broadly with "sins of the mouth," which would include blasphemy and the uttering of curses.[5]

Although this may be true enough, the English translation of the *Summa Theologica* leaves little doubt that eating, rather than drinking or swearing, was the main concern of Thomas's comments on gluttony. Referring frequently to Gregory I, Thomas defines gluttony as "inordinate concupiscence in eating." He defends the verse summarizing the forms of gluttony (hastily, sumptuously, too much, greedily, daintily), arguing that "inordinate concupiscence" may arise either in connection with the *type* of food eaten or the way in which it is consumed.[6] He explains that gluttony is a mortal sin in virtue of its capacity to give rise to further acts of sin.

> It is true that food itself is directed to something as its end: but since that end, namely the sustaining of life, is most desirable and whereas life cannot be sustained without food, it follows that food too is most desirable: indeed, nearly all the toil of man's life is directed thereto, according to Eccles. 6:7, "All the labor of man is for his mouth." Yet gluttony seems to be about pleasures of food rather than about food itself; wherefore, as Augustine says (De Vera Relig. liii), "with such food as is good for the worthless body, men desire to be fed," wherein namely the pleasure consists, "rather than to be filled: since the whole end of that desire is this—not to thirst and not to hunger."[7]

Gluttony is a mortal sin because the glutton places pleasure before nourishment. In doing so, gluttony induces one to other acts of mere self-gratification.[8] For present purposes, the way dietary rules are absorbed into larger questions about the moral justifiability of sensual pleasure is noteworthy. Seemingly trivial concerns about "eating daintily" or "too soon" point toward patently more meaningful questions about one's commitment to moral action in general. Yet *none* of these rationales display cognizance of any potential link between overconsumption of food and present-day worries over obesity or chronic disease.

This should not be taken to imply that the ancients saw no connection at all between diet and health. The ancient Greek school of medicine associated with Hippocrates of Cos (c. 460–370 BCE) is known to have recommended diet, exercise, and fresh air as components of both a daily regimen and for the treatment of disease. We should notice that these ancient and medieval writers saw no particular puzzle in linking a healthful and a moral practice. The connection between diet and health must have been understood in a spiritual sense as much as it was in a medical sense. If there is any carryover from these ancient views on diet for us today, it has been filtered through a mesh of scientific and philosophical developments that have dramatically altered the way that diet, health, and ethics are understood. In fact, all three of these ideas underwent dramatic transformation during sixteenth, seventeenth, and eighteenth centuries.

Dietetics in the Modern Era

So many different theories of healthful dietary practice were advanced during the three centuries that mark the rise of contemporary science that one questions whether *any* perspective can be characterized as the dominant one. Historian Stephen Shapin says that influential dietary advice from the pen of Luigi Cornaro (1467–1566) broke the connection between health and longevity, on the one hand, and religious or spiritual motivation, on the other. This lack of connection was also reinforced by Francis Bacon (1561–1626), who stressed dietary moderation as the route to long and happy life. Ordinary people were encouraged not to deny themselves any sense of pleasure from food or the occasional feast, while also being cautioned against the dangers of unconstrained excess.[9] Tristram Stuart's history of early modern dietary beliefs stresses vegetarianism. In Stuart's view, the dominant dietary paradigm of the era displayed a firm commitment to meat consumption in copious quantities. Against this view, a small but dedicated crew of reformers campaigned for plant-based diets. Their efforts were based on a blend of what we might today call animal ethics, reference to earlier ascetic texts, and a faith in these diets' healthfulness that was often based on personal experience. Stuart writes that pro-vegetarian views were actively repressed, accounting (perhaps) for a closeted vegetarian practice among such members of the intellectual elite as Sir Isaac Newton and his sometime physician, George Cheyne (1671–1743). The publication of Cheyne's *The English Malady* in 1733 made vegetarianism scientifically respectable, or at least discussable. Stuart notes that Cheyne's reportedly successful treatments of the "the English malady" involved first dosing patients with mercury. This treatment was generally followed by a worsening of their symptoms. Cheyne may not have known that mercury is a potent neurotoxin. When his originally melancholic patients (who may have been plagued by mild depression) began to display more dramatic forms of sensory impairment and lack of physical coordination (now associated with mercury poisoning), Cheyne would suspend the mercury treatment and recommend a plant-based diet. And indeed, they got better, providing (for Cheyne) confirmation of his dietary hypothesis.[10]

While views on diet were being more and more closely tied to health, the theme of moral *ascesis* that had been carried over into Christian views on gluttony became less and less important. Shapin denies the existence of any significant cultural shift between the ancient and modern world with respect to the moral significance of diet.[11] Nevertheless, his overview of the many writers from the modern era who urged dietary moderation on ethical grounds betrays an important difference. As opposed to ancient writers in search of the discipline needed to *achieve* virtue, the moderns emphasize the need to *display* a virtue already obtained. In short, gluttonous conduct has come to be seen as a visible indicator of bad character, and the modern concern is fixated on reputation rather than *ascesis*. Although there is a superficial continuity in moral discourse, it becomes possible for moderns to be skeptical about the link between dietary moderation and virtue in a way that would not have occurred to the Greeks.

In contradiction to his own claim of consistency between the ancient and modern world, Shapin cites passages from Bacon that point out the lack of evidence linking dietary discipline to longevity, along with Michel de Montaigne's skepticism of dietary moralizing.[12] Thus, although the tradition of associating diet with morality was continued in books on comportment and self-medication, it was eroding among those writers who were setting the stage for the theories of knowledge and ethics that would come to typify the modern era. The change may have reflected a shift in the modern conception of morality as much as it reflected a change in medicine or dietary science. In the ancient world, morality was articulated in terms of virtuous practice, reinforced and validated by the social environment in which the virtuous person lived. In this context, a dietary *ascesis* might be undertaken in much the same vein as weight training or long-distance running: it is simultaneously a practice of self-discipline and a preparation for trials to come. As the Renaissance morphed into the Enlightenment, morality came to be articulated in terms of two seemingly incompatible philosophies. On the one hand, morality was a matter of cultivating a good will, an internal ideal or intention guided by abstract principles of right. On the other hand, morality was a matter of acting in a manner that achieved the best outcome. Although much of nineteenth- and twentieth-century ethics would be devoted to the battle between these viewpoints, they shared some characteristics that may have been decisive for the way that diet and morality have come to be related in the present day.

First of all, the potential tension between emphasizing one's will or intention and emphasizing the outcome underscores the way that someone can do the right thing for the wrong reason. Acts of temperance or charity that are done for the sake of reputation betray a lack of the proper will, even when they achieve laudable outcomes. More subtly, whether the focus is on intentions or results, morality is understood to arise in connection with decision or choice—conscious reflection on what should be done. Routines and habits may *become* topics for morality when they are subjected to the kind of mental scrutiny that we associate with decision making, but modern ethical theorists did not look to a person's unreflective involvement with quotidian tasks when they wanted to understand moral behavior. Ethical questions are implicitly taken to presuppose a minimal level of importance or weight. While nothing precludes dietary questions from *rising* to a level of significance worthy of deliberative decision making, it has been more typical to regard someone's obsession with day-to-day questions of personal comportment (or my dinner companion's interest in weight loss) as a lack of the seriousness needed for questions of morality. Finally and most significantly, modern ethical theories emphasize a feature that philosophers refer to as the "other-regarding" aspects of action. While neither Kant nor British philosopher Jeremy Bentham (1748–1832) denied that self-regarding implications of choice are relevant, the moral dimension of choice is primarily focused on the way that action affects other people. For Bentham, the founder of utilitarianism, it is considering the *total* outcome, including consequences for others, that is required by ethics. For Kant, it is regarding others as mere means for accomplishing one's goals that marks action as unethical. If you are the only person affected by a dietary choice, the other-regarding aspects of diet are unlikely to be very significant at all.

As noted already, John Stuart Mill's essay *On Liberty* may be the key source for marking a turn in contemporary thinking on diet and health. Although Mill is particularly interested in protecting an individual's ability to decide for themselves without interference from the state, he is also concerned with when one can, with propriety, opine or intrude upon the conduct of another as a matter of common morals. The general rule is that so long as a person's conduct does not adversely affect other people, there is no basis for objecting to it, no matter how destructive the conduct may be for the person engaging in it. In a concluding chapter on "Applications," Mill specifically considers excessive behavior typical of the glutton (though he does not mention gluttony by name). Drunkenness is, by Mill's lights, not an offense rising to the level of moral significance. One oversteps one's bounds by interfering in the inebriate's pattern of consumption unless and until it causes something that is overtly harmful to others. Elsewhere Mill writes on the personal and social benefits of cultivating virtue and good taste, but given the analysis in *On Liberty*, overconsumption of food or drink would not be morally significant in the most important sense, that is, in bringing about harm to others.

By the twentieth century, moralists had adopted a convention of distinguishing between prudence and morality. Broadly defined, the distinction turns on other-regarding implications, though the "seriousness" of an outcome also plays a role. Everyone makes decisions that can in some sense be called better or worse but that either don't affect other people or do so in only trivial ways: how one spends leisure time, whether one gets enough exercise, where one takes a vacation, what one eats. These choices involve prudence. Prudential choices are better than imprudent ones, but prudent choices are not *morally* better and it would be a mistake to accuse someone who makes an imprudent choice of committing a morally bad act. Choices only become *morally* significant if other parties are involved. If diet affects one's health, better or worse dietary choices involve prudence but not morality. And by the beginning of the twentieth century, there was little doubt that diet does affect health. Antoine Lavoisier (1743–1794) established the oxidation of food in producing the body's heat. Jacob Moleschott (1822–1893) published his *Theory of Nutrition* in 1858, making the case for an approach to dietetics based solely on scientifically established principles of digestion, metabolism, growth, and development. A sequence of chemists and physiologists have continued to build the scientific understanding of food consumption and its connection to both normal and pathological bodily processes ever since.

This gradual turn of thought may or may not have reflected the way that most people understood the relationship between morality and matters of diet. In every culture, food practice was and still is thoroughly saturated with meanings and normative expectations. Parenting advice takes little notice of a distinction between prudence and morality, so the admonition to eat your vegetables is easily internalized as an ethical command. There are also more subtle cultural dimensions. Shapin notes, "Then as now, declining a proffered dish or drink might be taken as an act of social disengagement."[13] Nevertheless, dietetics disappears almost entirely from philosophical ethics by the twentieth century. When it does appear, as in Susan Bordo's critique of the way that male expectations have reinforced a female body image that encourages obsession with dieting,[14] or in Peter Singer's advocacy of vegetarianism, the message remains consistent with the division between prudence and morality. In Bordo's criticism there is a suggestion that women should resist unhealthful dieting, but unlike Epicurus, Seneca, or even Montaigne, the argument has nothing to do with personal virtue and everything to do with the repressive attitudes of others. For Singer, the extension of moral concern beyond the human species leads to a conventional ethical argument against practices that cause harm to others.

A New Twist

In sum, Western culture ended the twentieth century with the view that what you eat is your business and yours alone. If you make yourself sick or obese and die young, that's your choice. Dietary advice could be proffered widely, nonetheless. Official organs of government or science would offer pyramids and guidelines, while entrepreneurs and health faddists would offer one weight-loss gimmick after another. In either case, dietary recommendations were understood to have the status of prudential advice. People might form harsh opinions of obese people, to be sure. The imprudent fatty remained, in that sense, a target of moral condescension along with drunkards, spendthrifts, impulsive gamblers, and the drug addicted or downtrodden. Weakness of the will may have been thought to underlie all these conditions, but there was nonetheless a line to be drawn that prohibited interference with this behavior except by family members or close friends who had an implicit license to act on an afflicted person's behalf. Civil society could not intervene unless and until the behavior came to be seen as a social problem. While this threshold had been reached (at least in the view of many) for drugs, gambling, and some cases of alcohol abuse, obesity was a personal and not a social problem.

While such a view of obesity was unquestioned at the midpoint of the twentieth century, that view is precisely what was changing at the dawn of the new millennium. Two developments conspired to bring about the change. One was the science of degenerative disease. Obesity was defined in terms of the body mass index (BMI) and statistically correlated with cardiovascular ailments and the early onset of type 2 diabetes. By the year 2000, the rates of obesity and hospitalizations for heart conditions and the debilitating effects of diabetes were climbing in tandem. This might yet have been viewed simply as imprudence on the part of the unlucky individuals struck with these health problems but for the second development, the rise of health insurance and socialized medicine. State-run programs of health care for the impoverished and the elderly became the norm throughout the industrialized world, and in many countries these programs covered the entire population. This system means that one person is paying taxes to support care for others, and if one group—the obese—is utilizing those dollars disproportionately, then it becomes possible to see overconsumption of food as imposing costs on others. The situation is the same among private insurance pools provided by employers. Put succinctly, you are paying taxes or insurance premiums to pay for my diabetes medicine or my heart surgery. If I *caused* my own diabetes or heart disease through intemperate eating, it's no longer about my own prudence. Now I'm imposing costs on *you*.

Under such an arrangement, eating a healthy diet is a moral obligation in much the same way as recycling or limiting one's emission of greenhouse gases. It is not that these actions return any direct benefit to the person who does them. What is more, it is not even the case that any one person's *failure* to do them results in any significant harm. If only one or two people in a thousand are overweight, our health care systems are unaffected. But the collective impact of statistically observable increases in the rates of degenerative disease begins to have noticeable effects on our health care systems and our insurance premiums. However, as with obligations to recycle or limit emissions, the language of traditional morality is ill suited to articulating a strong claim of ethical responsibility. My failure to recycle plastic bottles or to abstain from a pleasure ride in a low-mileage vehicle does not in itself *cause* a harmful environmental outcome. Similarly, my obesity does not in itself *cause* the uptick in degenerative disease. Just as some skeptics have questioned the assertion that individuals can be said to have moral obligations to limit their greenhouse gas emissions,[15] the collective-harm analysis leaves something to be desired when it comes to explaining how diet and health can be conceptualized as an ethical issue. This has not deterred organizations from introducing incentives for employees or members to maintain healthy weights or from imposing penalties on those who do not.[16]

A group of Dutch philosophers has developed a systematic framework for analyzing the ethics of diet and health. Suggesting that an analysis of moral responsibility has to match the analysis of causal responsibility, they sketched three general models for explaining in causal terms the increase in the obesity rate in industrialized nations. Then they considered the ethical implications of each model. The first model stresses individual behavior and maintains that the obesity rate has increased because individuals within the population are behaving badly. They are eating more or exercising less than non-obese individuals. It is possible that they are both eating more *and* exercising less. The result of this change in behavior is the noted uptick in the statistical rate of obesity and the epidemics of diabetes and heart disease. This model is attractive in part because of its appeal to common sense, but it may beg the most relevant question: *why* are individuals in contemporary industrial societies eating more and/or exercising less? The second model attacks that problem directly, asserting that the epidemic of dietary disease is being caused by a change in social institutions or cultural factors. Leading candidates would be government policies or business practices that directly affect the foods that people eat. Other possibilities would be educational or communicative changes that, in turn, influence individual behavior. The final model stresses medical causes. This model is suggested by the very language in which the phenomenon in question is described: an *epidemic* of dietary-based disease.[17]

These three hypotheses are introduced in part to critique the science on which an epidemic in obesity-related diseases has been diagnosed. Some have argued that the alleged statistical uptick is something of a shell game, a manipulation of statistics intended to create the impression of a social crisis so that new technologies can be introduced and policies changed in ways that will benefit the health care and/or insurance industries.[18] Indeed, the perception of an epidemic has spawned both the introduction of foods processed to appeal to an image of health and a proliferation of scientific studies intended to link diets to "personal genomics"— the construction of dietary recommendations or possibly even specific processed foods tailored to an individual's genetic makeup. The Dutch group raises a number of epistemological, methodological, and ethical questions about these scientific and technological developments,[19] but here I will reconstruct their three hypotheses with an eye toward how each approach might be expanded into a normative position in food ethics.

Individual Causation and Individual Moral Responsibility

As noted, the suggestion that individual decision making has caused the epidemic of dietary disease enjoys a certain amount of support from common sense. Not only is it individuals who eat and subsequently become obese, they do so through a series of actions that can plausibly be described as choices. The features that allowed Mill to characterize drunkenness as a form of imprudent behavior that becomes morally significant only when it begins to affect others apply at least as well to conduct that might be called gluttonous. Although Mill's analysis may have been important for distinguishing imprudence from immorality, it does not absolve the imprudent actor from responsibility for harm done to the self. Insobriety continues to be cited as a case of *akrasia*, or weakness of

the will. When one acts against one's better judgment, one is causally responsible for the results. If one is harmed by one's imprudent
consumption of food *or* spirits, one has no one to blame but oneself. When *others* are harmed, the blame is morally significant.

One might go on from this common-sense observation to recognize that harm resulting from poor judgment in respect to consumption of spirits depends on social context. In nineteenth-century
England, the public inebriant may have offended bystanders and
become annoying, but this can be distinguished from the threats
and physical injury caused by the abusive drunk. It is the turn
to physical violence rather than the lack of sobriety as such that
becomes the target of moral judgment. Even a drunken rider in
the nineteenth century would have caused little harm because
his horse knew better. By the twentieth century, the ubiquity of
automobiles and other forms of potentially hazardous machinery
significantly multiplied the opportunities for inadvertent harm
from self-induced temporary incompetence. Someone who engages
in inebriating activity without taking the necessary precaution of
a designated driver will readily be singled out for moral criticism.
Perhaps the situation with gluttony is the same, though we are only
now coming to recognize its costs.

This reasoning may strike readers as initially plausible; nonetheless, it may be wise to exercise caution. A predisposition to heap
blame or disdain on the obese may be doing the work that a causal
explanation was supposed to do. A significant body of social science and public health research documents and analyzes stigmatization of obesity. Fat people tend to be judged as less capable, less
reliable, and less worthy of respect than otherwise similar people
of normal weight or who are underweight.[20] We—or some of us,
at least—seem disposed to think that obese people have caused
themselves to be fat. Then we go on to act as if subsequent social
ills associated with the prevalence of obesity can also be laid at the
individual fat person's feet. While the belief that obese people have
caused their own fatness is not implausible, the further inference to
social ills is an example of the logician's composition fallacy: a claim
true of the group is invalidly attributed to individual members of

the group. The cause of a *social* problem (the increase of diet-related disease) is mistaken for the cause of an *individual's* decline in their personal health. Notice that this is *not* the case with respect to the insobriety example. Drunken drivers unambiguously cause highly significant forms of harm to the person and property of innocent victims. In contrast, even if the *social* phenomenon of obesity is harmful, it is difficult to see how obese people cause any kind of material or physical harm to anyone other than themselves.

Yet if individual fat people cannot be held causally responsible for a social trend, it might still be the case that a change in individual decision making is the way to reverse it. Once again, it seems commonsensical to infer that *if* obese people would make the individual meal-by-meal decisions that constrain their intake of their calories, or *if* they would undertake a fitness program to correct their BMI, the high statistical rates of obesity would disappear. If obesity really *is* causing an increase in diabetes and respiratory or heart disease, then reducing obesity implies that the epidemic of obesity-related disease should disappear, as well. So even if there are logical problems in attributing a social trend to the action of this or that individual, perhaps better decision making by individuals would still reverse the trend. And if this trend needs to be reversed for ethically sound reasons, it seems reasonable to infer that the people whose decision making could reverse the trend are ethically obligated to make the right choices. Something resembling this kind of reasoning must lie behind the persistent and widespread tendency to place moral responsibility for the epidemic of degenerative disease with people who are too fat. Correlatively, if you are not fat, it is not your responsibility to do anything different from what you are already doing. Once again, there is a parallel between this reasoning and arguments imputing a responsibility to individuals to take action to avoid or reverse environmental harms.[21]

Reasoning that holds individuals causally and prudentially responsible for their own obesity might also suggest that stigmatization of overweight body types supports an ethically laudable outcome. When overweight individuals are ostracized or suffer social consequences for their body shape, they have incentives above and beyond their personal health to correct the eating and activity behaviors that lead to obesity. Such an outcome would be socially good even if the stigmatization is itself ethically problematic. Although social science research on the effectiveness of these incentives is equivocal, the claim that overweight individuals bear the moral responsibility for ending the obesity epidemic implies that incentivizing corrective behavior is a good thing. However, the socially good outcome does not in itself justify ostracism, insults, or other social conventions directed against overweight people. An extensive literature in philosophical ethics illustrates situations in which actions that produce good results are still considered unjustified because they fail to show adequate respect and consideration for the people at which they are directed.[22] *Justifying* ostracism of the obese is thus a project in food ethics that awaits further argument for those who might wish to attempt it.

Sociocultural Causation and the Search for Responsibility

Although the individual responsibility hypothesis provides a seemingly plausible way to assign responsibility for the obesity epidemic, it does nothing to explain why a larger proportion of the population is overweight today than in the past. If it is just common sense to think that individuals are personally responsible for regulating their weight, it is also common sense to determine why this is much more difficult to do now than it has been in the past. Much of the physical labor that once required an expenditure of calories and kept bodies trim has been replaced by machines. Recreational activities that would have done the same thing have declined, and sedentary pursuits like watching television or monitoring a computer-mediated social network have taken their place. Instead of walking, people will hop in their car for a quick trip to the store, and the distance covered by the average daily work commute has lengthened to the point that walking or bicycling is not an option for many. There are, in fact, many ways in which contemporary lifestyles undermine fitness.

At the same time, there have been significant changes in the diet of people in industrialized countries, and nowhere are these changes more evident than in the United States. Among the changes studied by obesity scholars are a general increase in calorie intake,[23] an increase in the consumption of meals eaten away from home,[24] an increase in the proportion of dietary calories from simple sugars such as sweetened beverages,[25] and a decrease in the consumption of fruits and vegetables.[26] Although common experience provides a ready explanation for why changes in physical activity are occurring, these changes in dietary practice are more opaque. Dr. David Kessler, former head of the US Food and Drug Administration (FDA), is one of the more notable voices offering an account. In short, Kessler blames the food industry. Kessler argues that chain restaurants and food manufacturers—the companies that make and market processed food in boxes, cans, and freezer containers—have engaged in decades-long research to determine the neurological triggers for appetite. They have been so successful in this work that they have now developed formulations that are virtually irresistible to consumers. The ordinary drive to make a profit has led these companies to compete with one another to develop foods with salty-sweet tastes and fatty-crunchy mouthfeel. The reasons people prefer these foods may lie deep in the evolutionary history of the human species, but as these foods have become ubiquitous in grocery stores and restaurants they have crowded out the more healthful alternatives that were more typical of human diets in the past.[27]

An alternative explanation suggests that people are less knowledgeable about what to eat than they might have been in the past. One version of the thesis asserts that parents are morally responsible for passing on this information and that the contemporary obesity epidemic is thus something for which parents are to blame.[28] Somewhat more broadly, contemporary lifestyles have "commodified" food consumption, turning an activity that once would have involved significant planning and thought into a transaction motivated by immediate gratification. The transition has occurred slowly, with the skills that allowed a peasant family to transform recently harvested vegetables, grains, milled flours, milk, meat, and eggs into a nourishing meal beginning to erode even in the nineteenth century. In the present day, many men and women lack the skills for cooking and choosing their foods.[29] This thesis conjoins with the idea that imparting nutritional knowledge is an educational responsibility. In fact, educational expenditures on nutrition-oriented curricula have grown significantly.[30]

More subtly, a scientific failure may be at work. Nutritional science has origins in animal husbandry departments, where the goal was to develop feed rations focused on marbling and meat quality, on the one hand, and on rapid growth at low input cost, on the other. Nutrition education grew out of programs in home economics and county extension offices that were also housed in agricultural research institutions. Nutritional studies were historically neglected by physicians and biomedical researchers.[31] Beginning in the 1970s, Joan Dye Gussow, a professor of nutrition education at Columbia Teachers College in New York, began to argue that a reductive research strategy focused on identifying chemical compounds that contribute to growth and immunity had led nutrition scientists to drop the ball. The effect of all this research was to provide the food industry with an ever increasing list of additives that could be introduced into foods during the manufacturing process, and to deprive those who might question this process of any science-based leg to stand on. Although the pattern of research in these agricultural labs led to grants and publications for the scientists and supplied the food industry marketing ploys and profit-making opportunities, Gussow argued that sound nutritional advice was better focused on eating a balanced diet of whole foods (i.e., foods that had not been industrially processed).[32]

Gussow's work anticipated the more recent work of Australian scholar Gyorgy Scrinis, who coined the term *nutritionism* as a rebuke to nutritional scientists' fascination with nutrients as opposed to food.[33] The critique of nutritional science has been popularized through the book *In Defense of Food* by Michael Pollan. Pollan has also drawn links to agricultural policy, arguing that subsidies for corn production in the American Midwest have

made processed foods extremely cheap. He takes particular note
of the development and widespread use of high fructose sweeteners made from corn. Pollan has made extensive use of Wendell
Berry's adage, "Eating is an agricultural act," to promote the idea
that a dysfunctional agricultural policy lies at the root of the
obesity epidemic. The argument holds that agricultural universities in bed with chemical companies and the food industry have
used money from the farm bill to support research that makes
food obscenely inexpensive but not particularly healthy on either
nutritional or environmental grounds.[34] As mentioned already,
the popularity of Pollan's 2004 book *The Omnivore's Dilemma* may
be a significant factor in the rising interest in food ethics.

In fact, none of the above listed possible social causes for an obesity epidemic are mutually exclusive. *All* of them could have contributed to the increase in diet-related diseases among Americans and
other citizens of industrialized countries. The plurality of possible
causes has several implications that are ethically significant. First
of all, multiple causes complicate the transition from knowing what
triggered the obesity epidemic to diagnosing moral responsibility for
the costs and harms associated with it. At a minimum, responsibility
becomes ambiguous. Second, the possibility of more than one causal
agent provides an opportunity for dissembling and finger-pointing.
When it becomes time to turn from causal hypotheses to identifying who bears the blame or the responsibility to take action, one
possible agent can always point to a different one. This observation
is especially pertinent to Kessler's claim that the food industry is
responsible. It is easy for food industry firms to commission studies
and mount campaigns emphasizing a decline in physical activity or a
failure in education. Finally, there is a phenomenon in morality that
when responsibility is widely shared, no one is actually held accountable. To the extent that multiple causes come into play, the food
industry escapes responsibility and dodges action that might be taken
against them.

As food industry firms resist taking any share of the blame,
the multiple causation hypothesis also starts to suggest a general
political failure. When no private agent can be held accountable,

our attention turns naturally to government as the actor that must assume responsibility to do something. Indeed, Pollan blames government inaction directly. Increasing physical activity, regulating the food industry, and shifting priorities for research and education could all involve government action of one sort or another. To be sure, government actions in multiple domains might be supported by a number of ethical and political perspectives. Utilitarian and Kantian moral theories provide reasons for government to promote human flourishing, and in the case of dietary disease management, the prediction of excessive health care costs buttresses the human-flourishing rationale with the opportunity to curtail cost increases that threaten existing institutions for maintaining people's access to health care. Of course, at the same time that these arguments present a case *for* government action, libertarian arguments provide well-established reasons to oppose it. Those who are *not* obese are taxed to do something that governments have no business doing in the first place (e.g., providing a benefit), and government intervention in the economic activities of private enterprise (e.g., the food industry) is anathema to those who advocate limiting government and relying on the market. In short, while the social causation hypotheses are at least as plausible as an individual responsibility argument, the prescriptive response leaves us mired in the big government/small government debate that has stymied effective policy on many contemporary political issues.

Medical Causation and Its Moral Implications

As put forward by Michiel Korthals, the lead investigator in the Dutch group discussed previously, the medical causation hypothesis was closely tied to the emergence of genomics. Korthals was investigating the idea that some subset of the human population was genetically predisposed to obesity and that this medical discovery would suggest technological responses ranging from the construction of personalized diets to the development of a "pill" that would counteract whatever protein was being produced or regulated by offending genes. Korthals was skeptical not only of the science that was pointing in the direction of these technical responses but also of the desirability of "medicalizing" the problem in a manner that would put moral responsibility in the hands of physicians and the medical establishment. It looked a lot like an opportunity for creating a new profit center for hospitals, clinics, and attending medical staff.[35]

● Kessler has also been critical of the "fat gene" hypothesis, though for different reasons. He points out that it actually does nothing to explain an *increase* in the rate of obesity. If genes disposing some subset of the population to obesity exist, there is every reason to suppose that they were as widely dispersed through the population in previous generations as they are today. Even if some percentage of the population—say 30 percent—carries a gene that makes them vulnerable to higher than average body weight, there would still need to be some environmental change to trigger an increased response to this vulnerability. Any medically identifiable and even treatable conditions that contribute to obesity are at most contributory causes. There is still a reason to look for what is different in the environment, and for Kessler this is the irresponsible predatory behavior of the food industry. A medical contributing cause might provide an alternative response to the obesity epidemic in the form of drugs, surgery, or a behavioral therapy. However, the question of whether this is the ethically *appropriate* response can only be decided after one has compared the benefits, risks, and costs of medical intervention with a policy-oriented approach that reverses the changed environmental conditions that triggered the genetic predisposition.

In fact, Kessler reviews quite a few medical studies that have probed neurological and developmental responses to the dietary changes (more fat and sugar) that he associates with the marketing strategies of food industry firms. Contrary to a straightforward genetically based predisposition, the studies that Kessler cites would have rather widespread effects throughout the human population, irrespective of genetic difference. In particular, Kessler discusses the chemistry of brain development in early childhood, citing suggestive (but not conclusive) studies. Children who eat sweet and fatty foods develop a lifelong disposition to eat sweet and fatty foods, and they lose the body's natural ability to "turn off" the neural signal for hunger.[36] This would explain a rapid increase in the rate of obesity. As people who had heavy exposure to rich, sweetened foods (such as soft drinks, snack chips, and sweetened sodas or fruit drinks) during childhood mature, their increased attraction to sweet and salty foods combines with a weakened ability to feel satiated. An entire generation that was the first target of the post-World War II food industry products during their childhood years became obese and is now experiencing the onslaught of dietary disease.

Kessler's summary of research on diet and brain development does not examine more profound implications of research on potentially heritable changes in gene expression and cellular function that do not involve underlying changes in DNA. Food intake contributes to the biochemical environment that exists within each cell of a person's body. This is the literal sense in which you are what you eat. But the biochemical environment inside a cell is the environment in which DNA functions. Changes in this environment thus have the potential to affect genes, both in ordinary cell activities and also during mitosis and meiosis. It is thus possible for a person's DNA to be primed, so to speak, for its cellular environment by biochemical events experienced in the reproductive cells of the mother or grandmother. Explanations that attribute the rise of dietary disease to epigenetic influences get around Kessler's observation that our DNA is pretty much like that previous generations. There are now hypotheses being formed and investigated which propose that changes in the diet of current generations' grandparents may have laid the foundations for an uptick in diet-related degenerative conditions such as cancer, heart disease, and diabetes.[37] If these empirical speculations are corroborated by future research, the import of nutritional genomics will lie less in the "medicalization" of obesity feared by Korthals and his

Dutch colleagues and more in a deep and systematic rethinking of the concept of moral responsibility within the context of obesity and public health.

There are two key ethical implications to draw from this summary of the medical causation hypothesis. First, it is important to once again emphasize that the various causal mechanisms linking diet, gene expression, neural activity, and brain or cellular development do not entail any logical conflict with the possibility that this current generation of obese people might have spent many hours watching television and engaging in sedentary work. Indeed, the "reduced physical activity" and the "changed diet" explanations are complementary rather competing hypotheses. Nevertheless, the presence of a complementary cause makes empirical corroboration of the complex hypotheses linking diet, obesity, and genetics difficult, especially in the absence of experimental studies on human children (studies that would never be approved by an ethics review board in virtue of their high level of risk).This gives libertarians and food industry advocates a great deal of room to sow the seeds of doubt. In short, the empirical debate over medical causation will occur in a context dominated by ideological viewpoints and economic interests. It is possible to portray Kessler himself as an enemy of private enterprise and an advocate of government intervention. Disentangling the ethics, the politics, and the science will prove to be daunting task for scholars in science studies for many years to come.

The second important thing to notice is that although reduced physical activity and changed diet are complementary empirical hypotheses, they suggest contradictory analyses of moral responsibility. If a man or woman is obese because he or she spends work and leisure time seated or reclined in front of a computer screen or television set, it is reasonable to hold that man or woman morally responsible for a disease contracted as a result of *being* obese. Social scientists and statisticians may caution us about inferring a causal connection at the individual level from data that track social trends, but people will continue to find fat, lazy people morally responsible for their own health problems as long as overconsumption or the reduced-activity explanation continues to be persuasive. However, if you are diabetic or overweight because of what your grandmother ate, *she* is the one who is responsible for your health problems. She may not be morally culpable, but that is because she couldn't possibly have known what she was doing. The important lesson is there is no clear sense in which the person who has inherited the predisposition for dietary disease can be held fully responsible when the disease actually develops.

Looking Ahead

We have seen that there is in fact a long history of moral dietetics, but that there is also a cataclysmic break that separates traditional modes of moral and scientific thinking on diet and obesity from our present-day circumstances. People living through the statistically observable rise in obesity and the corresponding increase in diabetes, heart disease, and certain cancers lack a clear or noncontroversial way in which to characterize dietary practice as a moral obligation. They do not lack personal incentives to be thin or healthy: the ample scientific evidence supports the common-sense observation that fat people are less successful in a variety of life pursuits. But it is difficult to move beyond this purely prudential description of the reason to control one's body weight and to characterize it as a moral obligation.

One possible formulation draws an analogy with collective action or common pool resource dilemmas. According to this view, being overweight imposes an unjust cost on the non-obese people in one's insurance pool in much the same way that consumption activities contribute to pollution or climate change. The collective result of many individual actions contributes to a burden shared by all, even though the contribution to this collective result that is made by any individual would be insufficient to cause harm in the absence of similar activity by others. This analogy is imperfect. Just as everyone suffers from air or water pollution, everyone pays the higher rates for health insurance and suffers from the strain that dietary diseases place on the health care system. But

the fact that obese people are the ones *suffering* from these diseases and the fact that as individuals they have strong prudential
incentives to avoid them marks an important difference. Indeed,
to tell an obese person that they are morally responsible to keep
insurance premiums down sounds rather like blaming the victims
of this increase in the rates of dietary disease. Few obese people
relish their condition, and empirical studies on the ineffective-
ness of diets provides even more reason to question whether the
collective-result-of-individual-action account of moral responsibility provides a reasonable or effective way to characterize the relevant moral responsibility.

Recent work in diet-health relationships that emphasizes brain
development, genomics, and epigenetic mechanisms provides a
further basis on which to question any approach that holds individuals wholly responsibility for the increase in obesity or dietary disease. Medical authorities such as David Kessler are thus pointing
increasingly to social causes, and recommending social remedies.
But these remedies conflict deeply with the same philosophical traditions that have led us to think that what we eat is nobody's business but our own. These traditions also draw lines of responsibility
that separate the appropriate sphere for government-sponsored
activity from a sphere in which actors are free to develop and market products based solely on consumer demand. Here again, if
increasing rates of dietary disease bear some resemblance to environmental issues and collective action dilemmas, government regulation may be viewed as quite appropriate and politically justified.
But the role of governments in regulating pollution or environmental impact has hardly been noncontroversial, and the differences
between pollution cases and the uptick in dietary disease may well
prove more persuasive.

The upshot, then, is that we lack an effective moral vocabulary
in which to even discuss or debate the questions of diet and obesity.
The philosophical implication of *this* observation will depend on
one's meta-ethical views. Philosophers who hold that the basic concepts and underlying logic of ethics is a matter of analytic truths and
noncontradiction must either conclude that there is really nothing

for ethicists to say about diet and obesity, or that my discussion in this chapter has overlooked some crucial observation that will resolve the problem and show us where the ethical responsibility really lies. I am more inclined to think that our moral capacities are a function of our actual discourse. If we begin to discuss and debate the ethical implications of diet and obesity, we will discover and develop more powerful and more persuasive ways in which to specify and articulate the norms and principles that will allow us move forward in addressing a profound and difficult social problem that is causing pain and misery for many individuals. More broadly still, the ethical malaise that shrouds the diet and obesity issue like a fog might begin to lift a bit if we begin to see the significance of food as an intersectional locus, as a point of contact that integrates seemingly far-flung and unconnected social and political topics to our personal lives.

What is more, this conversational thesis would be strengthened by an intersectional approach to food ethics. We need to talk about the ethics of diet and obesity in part simply to achieve a social recognition that there is a deep tension here: when we think about the *causes* of obesity and degenerative disease we find many candidates, all of which could easily be operating simultaneously and contributing interactively to our mounting social problem. But when we shift to a discussion of diet and obesity as a matter of ethics we find ourselves with a number of mutually exclusive possibilities: it's *either* individuals or society at large (or specific groups such as government and food industry firms) or our genes that get characterized as morally responsible, but adopting any one hypothesis gets everyone else off the hook! This doesn't really make sense, but as we shall see in later chapters, neither does thinking that obesity, food security, and impact on nonhumans and the environment are wholly separate matters. Food ethics lives in the relations, and it is appropriate to put more connection points on the table.

Chapter 4

The Fundamental Problem
of Food Ethics

The obesity crisis is a problem of too much food, but we are much more accustomed to thinking of food ethics as a problem of not enough. Certainly everyone has some appreciation of the discomfort that accompanies hunger, even if it has only been experienced waiting for dinner to be served. To be sure, it would be grotesquely inappropriate to equate the anxiety of temporary hunger pangs with the suffering of people who do without sufficient food for long periods of time. If everyone has some limited personal acquaintance with the sensation of being hungry, surely everyone has also had some indirect exposure to the seriousness of extreme or chronic food deprivation as it is experienced by poor people around the globe. The idea that people who are *not* hungry have some kind of moral duty to attend to the needs of those who are dates back to biblical times, if not further.

Since the end of World War II and the creation of global institutions for governance and development, a somewhat specialized vocabulary has emerged for discussing the problem of hunger. As distinct from the wealthy person undertaking a voluntary fast, a person who is enduring extended deprivation of food is said to be "undernourished" or "food insecure." The term *food security* is felicitous in that it can naturally be understood to include both people who are presently enduring a condition of undernourishment and also people who cannot feel confident that they will have adequate food in the immediate future. Colloquially, such people "don't know where their next meal is coming from." The evils associated with food deprivation and undernourishment are legion. Beyond the physical sensation of hunger and the suffering associated with it, undernourished people are vulnerable to a host of diseases and chronic ailments. Undernourished children are apt to suffer permanent damage from stunting of growth and interference in cognitive development. Although I will emphasize the ethical dimensions of food security in this chapter, it is important to fix in one's mind both the seriousness and the breadth of afflictions that besiege hungry people.

Hunger is a problem everywhere. Industrialized societies have developed a variety of social programs to prevent chronic undernourishment and to address even short-term food deprivation that might be experienced in the wake of an economic crisis or a natural disaster. Nevertheless, various forms of food insecurity persist. For example, a document prepared for the US Department of Agriculture (USDA) states, "An estimated 14.5% of American households were food insecure at least some time during the year in 2012, meaning they lacked access to enough food for an active, healthy life for all household members."[1] According to the Food and Agriculture Organization (FAO) of the United Nations, approximately 15 percent of the population in less-developed countries experience more extended periods of chronic hunger and undernourishment. For this group, hunger is truly a life-threatening experience.[2] It is important to have some grasp of the facts and figures indicating the extent and nature of hunger in both industrialized and less-developed nations, and readers can obtain a more complete and up-to-date picture with a simple Internet search. The task here is to focus on the ethics of food security.

Hunger as a Moral Problem

Being hungry is not a good thing. You did not need to open a book on food ethics to know that. Although a few readers may have been unlucky enough to experience true hunger, I suspect that many people approach the moral problem of hunger by reflecting on circumstances in which someone has made an appeal to them for aid. You may have been accosted on the street by someone asking for pocket change to buy something to eat. In some cities around the globe, you can be surrounded by beggars, obviously poor and needy. You may also have been solicited for formal appeals through your church or school. Seemingly hundreds of charitable organizations make such appeals, ranging from Care, Oxfam, or Unicef—all international organizations working for poverty relief—to a local food bank. You may have been asked to contribute a can or bag of food when you pay your grocery bill at your local market. Although the feeling that you *should* make these donations up to the limit of your ability to do so is far from universal, my suspicion is that many people start out with an untutored intuitive sense that doing something to help hungry people is a morally good thing to do.

But unlike the problem of diet and obesity, there has been a lot written, especially in the last fifty years, on the moral dimensions of hunger. The legacy of work by the "3 Ms" of nineteenth-century philosophy (Malthus, Marx, and Mill) got a significant boost with the publication of Peter Singer's paper "Famine, Affluence and Morality" in 1972. Singer was making the case for individuals to make voluntary contributions toward relief of a then raging famine in East Bengal, India. His argument was simple: if you can do something to prevent a serious evil for someone at comparatively little cost to yourself, you should do so. He claimed that virtually any theory of ethics would support this principle, and noted that even a rather paltry contribution of a few dollars would mean an enormous amount for the famine victims, if everyone (or almost everyone) would follow suit.[3]

When Singer's article appeared, there was a widespread fear that food security would be getting worse. A 1968 book by Paul and Anne Ehrlich (only Paul was given authorship credit) predicted a wave of food shortages throughout Asia, Africa, and Latin America throughout the 1970s.[4] The ecologist Garrett Hardin had produced a series of articles arguing that uncontrolled population growth made a wave of famines and violent conflict over food inevitable. The only ethical response, said Hardin, was to recognize that the number of deaths due to starvation and the diseases of undernourishment would ultimately be lower than the death toll from unabated population growth. Altruism—offering aid of the form that Singer was recommending—was an ethically short-sighted course of action, in Hardin's view.[5] Hardin's argument may appeal to those who do *not* feel that tug on the heartstrings when they are approached by someone soliciting help for food. At a minimum, it suggests that there may be more to some of these stories than first meets the eye.

Many college courses have included articles from Singer and Hardin to suggest the basic starting points for a discussion of ethics and food security. As time has passed, the sense of impending and recurrent famines has waned, however. Writing in 2009, the Ehrlichs admit that they were too pessimistic. They credit Green Revolution-style agricultural development projects (discussed in Chapter 7) with averting the famines that they predicted for the 1970s. They also express the view that the fundamental problems have yet to be addressed.[6] To a considerable extent, the overarching moral issues of resource scarcity and population growth have now been subsumed in a larger debate over environmental sustainability—a topic that we will defer to Chapter 6. But this does not mean that the topic of food security has gone away. If Hardin and the Ehrlichs were wrong to emphasize widespread famine, the statistics summarized above show that undernourishment is still a problem for a significant percentage of the world's poor.

Meanwhile, much of the philosophical thinking in response to Singer's article strayed from its initial focus on the problem of food security. As noted in Chapter 1, philosophers such as Peter Unger stressed the idea that if one truly followed Singer's logic, one would feel compelled to give much more than we typically do. While Unger endorsed this view, other philosophers have contested it, and Singer himself has admitted that it may simply be too burdensome from a psychological perspective.[7] Others have focused on his suggestion that it is morally irrelevant whether the hungry person is right in front of us or in a distant land. The debate on this point has taken up whether it is morally acceptable to give priority to people in one's family, in one's neighborhood, in one's city or nation. These debates truly belong in food ethics, but they do nothing to help us think more critically about food security. One problem is that they leave our intuition that the moral dimensions of hunger are fully captured by that experience of a needy person asking us for help with their next meal intact.

Amartya Sen on Poverty and Famine

The 1970s, 1980s, and 1990s saw relatively few famines aside from those associated with war or conflict. The problem of food security began to be seen less as a crisis situation (though food crises do occur) and more as a problem of chronic poverty. One of the main reasons for this shift in thinking was an influential analysis of two famines by Nobel Prize-winning economist Amartya Sen. Using a variety of data sources, Sen argued that in the East Bengal and Ethiopian famines, significant supplies of food were either locally available or could have been sourced from nearby. Sen's work showed that in these cases, at least, food insecurity was not a problem of resource scarcity. This finding ran directly counter to the thinking that was being promulgated by Hardin and the Ehrlichs. As Sen's work began to sink in among those working on food security, it began to be summarized in slogan-like terms: hunger is not a problem of too little production; it's an imbalance in food distribution.

But there was much more to Sen's analysis than this. In *Poverty and Famines: An Essay on Entitlement and Deprivation*, Sen proposed the idea that we should think of food security in terms of an individual's "food entitlement." The word *entitlement* is potentially misleading. Sen was not thinking of a legal entitlement (such as a health benefit or employment insurance) or even a moral "right to food" that could be demanded from some person or government. Sen used this term to describe three different ways in which the security of a given individual's access to food was institutionally structured. First is the *income entitlement*: you are food secure if your income is sufficient for you to go out and purchase food. For most people in the industrialized world, food security is structured as an income entitlement. We buy our food at grocery stores and restaurants. Accordingly, we become food insecure when our income does not enable us to purchase enough food to satisfy our needs.

Again, most industrialized societies have instituted some kind of backup for situations where individuals become food insecure. The United States has its Supplemental Nutrition Assistance Program (SNAP), the successor to food stamps. People on reduced incomes or who have lost their jobs apply for this program and receive vouchers that can only be exchanged for food. In 2012, the United Kingdom announced that similar programs would be made available on a local basis. Other countries have addressed the problem by creating general income assistance programs that assure a level of income adequate for meeting basic needs, including food and shelter. There are also nongovernmental charitable programs such as soup kitchens that provide all these different types of access to food for people whose income entitlement is inadequate. Sen's term for this type of institutional structure is a *gift or grant entitlement.*

The East Bengal food shortages that Sen analyzed in his book could be explained in terms of rapid inflation that created extreme vulnerability for poor people with limited incomes. Due to economic events having nothing to do with the food sector, prices soared, and though food was in the shops, those whose income had not grown with price inflation simply could not afford it. Anyone with a fixed amount voucher such as the SNAP benefit would not have been able to afford it either, though such programs were not available in India during the time Sen studied. His point was that even with no disruption in the food supply, economic volatility could threaten the security of someone with an income-based food entitlement. But this did not explain the Ethiopian famines that Sen studied. These famines had occurred because of drought and the pestilence that often accompanies a dramatic shift in the weather. Although food was available not far from the region where the famine occurred, it was not made available to the Ethiopian farmers whose crops had been lost.

These farmers had what Sen called a *direct production entitle-ment*. They were growing most of what they ate. Ironically, small farmers with a direct production entitlement *do not* suffer from the kind of food insecurity that struck in East Bengal. They might suffer during times of economic upheaval, but as long as their access to land remains secure, they will still have enough to eat. Or perhaps we should say, as long as their access to land, seeds, fertilizing nutrients (which may come from animal manures), and water is secure, they have enough to eat. People with a direct production entitlement are vulnerable to the classic natural causes of famine: a plague of locusts, a drought, a flood, a dust storm, or a hurricane. They are also vulnerable to interlopers who steal their crop right from the field where it is growing. Sometimes these interlopers are wild animals—deer, antelope, or insects—sometimes they are other human beings.[8]

This three-pronged structure of food entitlements—the income-based entitlement, the gift or grant entitlement, and the direct production entitlement—lays the basis for a more sophisticated understanding of food security. It allows us to go beyond the "it's a problem of distribution" nostrum. Recognizing that each type of entitlement has its own peculiar form of vulnerability puts us in a position to think much more carefully and insightfully about the ethics of hunger and malnourishment. Most importantly, it allows us to see that these different types of entitlement are interlinked in unexpected and even nefarious ways.

Hunger as a Moral Problem Redux

Many philosophers who read Sen's work on famine or his later work on capabilities and human development took a somewhat oversimplified message from his work. Food security, they thought, is not a problem of too little food. Rather, it is a problem associated with poverty and lack of income. If we can get people out of poverty, the problem of food security will go away. The key target is the World Bank standard for *extreme poverty*, which at this writing is one euro (€1) per day (or approximately $475 per year) in income.

It is people *below* this threshold who are at the greatest risk for undernutrition and food insecurity. Given this assumption, it is not too surprising that philosophers who write on global ethics or the ethics of development had pretty much stopped talking about hunger, food security, or the right to food by 2005.[9] Everyone understands that people need food, but they need shelter, clothing, and health care, too. If extreme poverty is the source of their inability to meet all of these needs, there's little point in belaboring the ethics of food.

In fact, much of the current work in global ethics follows Peter Singer's original 1972 article in presuming that the big job is convincing well-off people that they should be doing more to help the poor. But in this connection, it will be illuminating to go back to the 1970s before Sen's work on famine had its influence. When people read Singer's original article, everyone assumed that the way to help these famine victims was to send them food. Indeed, a great deal of global development assistance in the 1970s was organized around gifts of food. Starting in 1954, the United States organized a major component of its foreign aid program through Public Law (PL) 480, known popularly as the "Food for Peace" program. US farm surplus would be purchased by the government and redistributed to hungry people around the globe. Food for Peace was (and continues to be) one of the few foreign aid programs enjoying the support of most Americans.

Feeding hungry people became the primary message for all manner of charitable giving intended to address poverty in underdeveloped parts of the world. This is a theme that continues today, and the idea that your contribution to Oxfam or World Vision is helping to feed hungry children may well be a major part of your motivation for digging out your wallet and throwing some cash in the direction of these charities. There is thus a sense in which Singer's original argument continues to be a focal point of the ethics of hunger. When Sen's analysis showed that there was probably some food already there, that took some of the fire out of the charitable appeal. But a careful reading of his food entitlements work should have suggested something more complex.

When food aid shows up on the doorstep of an impoverished nation, it surely does benefit the people with income-based entitlements. Whether it is literally given away or sold (as it more typically is) at a concessional below-market price, having this kind of food available in local markets of the underdeveloped world is a good thing for the urban poor. But simple economics suggests that it is also going to depress the price of food that was already there. After all, if 15 or 20 percent of the people who were in the market for food now have their needs satisfied elsewhere, sellers are going to reduce their prices in order to clear inventories. This is especially true for goods that are perishable. And most food that has not been processed is highly perishable.

But what is happening to the people with direct production entitlements? What is happening to farmers who also may be living on one euro per day? Although subsistence farmers are producing most of what they eat, they have other needs as well. They fulfill these needs by loading some of their beans, millet, or barley into a basket and hauling it down to the village market. They eat what they produce, but they also need to trade some of what they produce in order to meet their other basic needs. They are the ones sitting there with food to sell when the food aid truck rolls into town with big bags of corn, rice, wheat, or soy labeled with "Food for Peace" and American flags. They are the ones whose faces fall when they realize that they will not be able to sell their beans, millet, and barley for the price that they expected. In short, the practice of extending charitable assistance to the urban poor (or supplementing their income-based food entitlement with a gift or grant) actually undercuts the livelihoods of people with direct production food entitlements.[10] And that *is* the fundamental problem in food ethics.

Food Security for the Farming Poor

Strengthening food security for the urban poor has traditionally been thought of as a two-pronged problem: raise incomes, but keep food prices low. There is good common sense at work here, and it is hard to quarrel with it as long as one is thinking about food security in terms of income-based entitlements or gifts and grants. But the complexities of food security for poor farmers continue well beyond the situation of a poor farmer (often a woman, by the way) sitting behind a large basket of lentils at a local market watching her prospects plummet when the village food aid arrives. The first thing to notice has been mentioned already. *All* these farmers need to sell something in order to survive. There may be a few people who have truly escaped the need to trade some of the agricultural goods that they grow in order to get access to tools, medicines, and other basic needs, but that number is vanishingly small. What is more, even subsistence farmers who eat what they grow are very likely to be in the market for food themselves occasionally. There will lean times of the year, for example, when one's crop is in the ground and whatever was put by in last year's harvest is starting to run short. At that moment, these farmers may wish that food was cheap, but it is important to remember that what money they have to buy food in the lean times is a function of what they were able to get for the food they sold some time earlier.

A second point concerns the very idea of "subsistence farmers." The first image that comes to mind is people growing literally everything they eat, and eating everything they grow. There *are* farmers who approximate that picture, but it is more typical for farmers to be growing one crop (or one part of the crop) for household consumption and something else to generate cash income. This is especially the case for smallholders who are growing something that they *cannot* eat: cotton, coffee, tea, or oil crops like palm or soy. Very few such farmers are growing *only* their cash crop. Worldwide, they will have small plots, a tree or two, garden boxes, or even flowerpots growing something they can eat. They might also scavenge in the forest for fruits, mushrooms, or tubers. And they may also fish and hunt. It is hard for people who buy everything they eat to imagine all the ways in which a direct production food entitlement might be structured, but it is going to look very different from one farm to the next, let alone one continent to the next. Seasonality affects the availability of direct food entitlements in diverse ways, so the food security of a small farmer is a fairly complex blend of being able to produce something to eat, being able to produce something to sell, and finally being able to buy whatever is needed to supplement during the rest of the year.

A third point involves the number of people affected by vulnerability in production-based entitlements. United Nations special rapporteur for food security Olivier de Schutter estimates that fully 50 percent of the global population living on less than one euro (roughly $1.30) per day are food producers. Another 20 percent of the extreme poor depend on incomes that are derived from the food and fiber production economy. Their food entitlement may be a blend of income and direct production, but their income depends on agriculture either as labor for the production of crops and livestock or indirectly through transport, supply, and other activities that support agriculture. An additional 10 percent are hunters or scavengers whose food entitlement depends at least partially on access to a forest or other common-pool resource. These percentages mean that classic forms of food aid benefit 20 percent of the world's poorest people while potentially harming the other 80 percent. It is in one sense perfectly correct to say that improving incomes is important for this 80 percent of the poorest of the poor, just as it is for the remaining 20 percent of the world's most poor living in cities. Poor farmers need larger and more secure sources of income in order to secure a wide range of their basic needs. However, the relationship between their ability to have a secure food entitlement and other capabilities that would be enhanced by increasing their income is rather more complex than that of the urban poor.

Economists and other experts in agricultural development debate what actually helps smallholders. Some argue that since smallholders do need to buy food (and especially since they are likely to buy it at the moment when prices are highest), they—like urban consumers—are actually going to benefit when global food prices are low. The way to get them out of poverty is to improve their technology, to enable them to have more to sell. Others argue that subsistence farmers need *two* things. They do need to improve the productivity of their farming operations, but they also need a slow and stable rise in the global price for agricultural commodities of all kinds. On this view, the food security of a smallholding subsistence farmer is increased even when there is an increase in the global price of cotton. I have been somewhat persuaded of this latter view by the work of Marcel Mazoyer and Laurence Roudart. In a magisterial treatment of agro-ecosystems throughout history, Mazoyer and Roudart demonstrate how farming systems enable and interpenetrate social institutions, anchoring a civilization in its ecological niche. They argue that the basic biological productivity—the ratio of energy expended and then reaped in the form of a harvestable crop—is simply inadequate for many of the poorest small farmers today. They need a better way. Yet at the same time, some smallholders who achieve significant levels of biological productivity remain in poverty because world food prices are just too low.[11]

One thing that makes economists leery of increasing global food prices is that small farmers are seldom the beneficiaries of increases. In fact, technical improvements in farm production methods cause a treadmill phenomenon that often harms smallholders. When a new technology becomes available, early adopters have lower than average production costs. Since the market price reflects average production costs, the early adopters reap windfall profits, which they typically invest in even greater production. As the technology becomes widely adopted, average production cost falls, and the market price falls with it. But late adopters then have higher than average production costs and must sell at a loss. Eventually, they are bankrupted and their farms are abandoned. Where land markets are stable, the land is purchased by early adopters with money from the windfall profits they earned before prices fell. When the new equilibrium is reached, all farmers find themselves running harder (i.e., producing more) to stay in the same place. They must produce and sell more of the crop to achieve the same level of economic well-being they enjoyed before the new technology became available.

Since the poorest farmers are also often the most disadvantaged in terms of their ability to adopt a new technology quickly, they are likely to wind up as the losers whenever the treadmill effect is observed. Meanwhile, larger or richer farmers snap up their lands and become richer still. But the economic perversity does not stop there. Since urban consumers are used to paying a set price for food, the drop in commodity prices at the farm level gives middlemen—grain traders, food processors, and retailers—an ideal opportunity to increase *their* profit margins. Too often, the benefit of more efficient, more productive farming methods is neither enjoyed by the poorest farmers nor passed on to the poorest people with income-based entitlements. However, we are not going to settle a debate among development specialists in an introduction to food ethics, and it may be best to simply leave this matter unresolved but with one qualification. *Everyone* agrees that price volatility is going to be bad for small farmers, who are likely to be selling when prices are low and buying when they are high. If a relative increase in the price of food is a good thing for poor farmers, this is still not an argument for the swings and dips that have been typical of the past decade.

An Ethics Critique

So what sense should we make of this economic picture in terms of ethics? In the chapter "Morality and the Myth of Scarcity" from *The Ethics of Aid and Trade*, I argue that both rights-based and utilitarian moral theories have framed problems in distributive justice through a lens which presumes a natural scarcity of goods.[12] That scarcity provides a clue to the way that income-based and production-based entitlements are in constant tension with one another. But unlike many other consumable goods, food needs can also be satiated. While it seems that people cannot get enough clothes or electronic gizmos, once they have enough to eat, well, that's enough. This creates a ceiling on the marketability of food. Once that ceiling is reached, farmers are simply unable to find buyers at any price.[13] When not forced to sell by their own poverty, farmers have generally learned to cope with this problem by withholding some of what they produce from the market. Sometimes they literally plow it under for use as next year's fertilizer. As noted, Sen's work shows that in recent times, hunger and famine are rarely associated with an absolute scarcity of consumable calories. Breakdowns in markets, property rights, or macroeconomic forces (such as rapid inflation) have been at the core of people's inability to have secure food entitlements, and the breakdowns can occur both when food cannot be bought *and* when it cannot be sold.

Second, the rights-oriented and utilitarian moral philosophies of the modern era do a great deal to legitimate the system of property rights that gives rise to the collapse of food entitlements on a local basis. Not only do farmers plow the crops they own into the ground while hungry people stand in food lines, they feel totally justified in doing so.[14] Indeed, some obvious extensions of these moral philosophies have proven to be quite insensitive to the vicissitudes faced by poor farmers. Utilitarians in particular have been tempted to deny that there is any problem at all. Something similar to the treadmill effect described above can be observed in many industries. As production or distribution methods evolve and achieve greater efficiencies, laggard firms go out of business. Although there is certainly pain and loss associated with these bankruptcies, the benefits elsewhere in the economy—if not to consumers in the form of lower prices then to the owners and employers of middle-man firms—outweigh those losses. This result is a fairly straight-forward potential Pareto improvement: the benefits to society as a whole are large enough to compensate for the harms that might be experienced by a few. This in turn is a plausible way to understand how the utilitarian maxim should be applied to food security. If we should do that thing that achieves the greatest good for the greatest number, we are at least moving in the right direction when benefits are sufficient to offset the costs.[15]

Indeed, growth-oriented economists such as Jeffrey Sachs have explicitly made this argument in defending the industrialization of agriculture in the developed world. They see this path, with constant concentration of ownership among farm producers and a continues flow of workers exiting agricultural employment, as the one that developing countries should take as well.[16] To these growth-oriented economists, agriculture is just another sector of the industrial economy. It should produce its goods as efficiently as possible. When the entire economy is organized according to this principle, we reach the utilitarian's goal of maximal possible welfare.[17] The suggestion that harms borne by those who lose their farming businesses are offset by economic growth brings this feature of the utilitarian philosophy to the forefront. Moral philosophers of a Kantian persuasion argue that this reasoning appears to treat the harm suffered by losers as a kind of collateral damage and thus violates Kant's master principle of morality, the categorical imperative. On this view, the moral rationale for insisting that costs be internalized has less to do with achieving true efficiencies and more to do with a set of side constraints that specify the ground rules for all efficiency-maximizing calculations. One must never act in a manner that treats other people solely as a means for achieving one's end. One must respect them as autonomous agents, which means that one must not compromise their basic freedom to pursue life plans that are themselves consistent with a principle of equal liberty for all.

The arguments for and against rationalizing the treadmill effects thus exemplify the great philosophical disagreement between outcome-oriented consequentialists and rights theorists. They also create an opening into rich and ongoing philosophical debates about the nature of property rights and economic risk-taking. The summary just given here is at best the first move in a philosophical debate between utilitarians and Kantians, who might continue to challenge one another on more finely tuned questions. For example, one might question whether extending the categorical imperative to cover economic losses is always justified even from a Kantian perspective. Yet viewing this problem simply as an instance of a deep philosophical quandary can miss the way that facts on the ground matter morally. The fact that it is agricultural production and food consumption that are at stake really matters, as does the fact that more than half of the poorest people in the world currently achieve their food capabilities through a direct production entitlement. Things might well stand differently for many other tensions between producers and consumers that arise in an industrial economy.

Indeed, one can find support for the idea that the case at hand is *not* like that of other exchange relations discussed by authors who have written on food entitlements from a Kantian perspective. Henry Shue's work on basic rights, for example, provides an argument for regarding food entitlements as having priority over many other political liberties and social entitlements. Shue argues that it would never be appropriate to expend social resources to secure even such liberties as freedom of speech or assembly until the right to food had been secured for all.[18] His reasoning relies on the idea that a right to free speech is not meaningful to someone who does not have enough to eat, so subsistence rights take priority over a great deal of other rights that are discussed in political theory. A number of similar arguments can be adduced for supporting a general right to food as recognized by the Universal Declaration of Human Rights. But this rights-oriented pattern of thinking provides far more support for the moral significance of a food consumer's side of the tension between the rural and urban poor than for producers. Smallholders who lose farms to bankruptcy are entitled to eat, but it is difficult to see how the right to food also entitles them to a price that allows them to recoup the costs of food production and continue to subsist as farmers. To put it another way, an argument for the right to food protects the rights of smallholders as persons, but it does nothing to suggest that they have any moral standing as farmers.

Why Farmers Matter Morally

One could, with logical consistency, end the matter here by concluding that smallholders qua farmers deserve no special consideration, though qua poor or hungry, they do. One can start the process of building an alternative analysis by noting that this is at least an ironic position. People who have the means to feed themselves and others are deprived of any opportunity to deploy

those means because the economic return on farming is insufficient to lift them from poverty. We then accept that they can be
placed in *total* receivership, dependent on gifts and grants in order
to eat food that they might well have produced themselves, and
we consider this to be a morally just result! As noted already, the
traditional utilitarian position has been able to embrace this irony
by emphasizing the compensating social benefits of more efficient
agricultural production.

A more sensitive ethics of poverty and hunger could begin by
remembering two points that have already been introduced. First,
less than half of the poor and hungry are visible in the streets, riot-
ing and clamoring when global food prices rise. The majority are
in the countryside where cameras seldom reach and, depending on
how they are situated with respect to complex markets for cotton,
wool, coffee, and tea, may actually be enjoying a modest boost to
the security of their capabilities and well-being. Poverty is not a
homogenous phenomenon. I am not suggesting that we should be
deaf to the cries of the urban poor. I am only saying that people who
depend on an income-based food entitlement are at best half of the
problem (and likely much less than half). Like many economists
who have addressed hunger on the ground, Sen's work is quite sensitive to the different ways that entitlements manifest themselves,
and also conflict with each other. Yet overgeneralizing the way that
food entitlements contribute to human development occludes our
understanding of poverty rather than enlightening it. We *should
not* deny farmers moral standing *as farmers*.

Second, the suggestion that hunger is "not a matter of production, but of distribution" can easily convey a result that is very damaging to the rural poor. The worst case occurs when well-meaning
people in the industrial world presume that a "distribution problem" simply means that they should redistribute some of their
own food surplus. Although there are a few occasions when this is
exactly what they should do, it is more typical for such efforts to
harm the rural poor, as discussed above. The more complicated case
involves the nature of development itself. Do we want a global agriculture that looks like that of the United States, where less than

2 percent of the population are farmers? Even if we do conclude in favor of growing the entire world out of this problem by ending smallholder production, it is ethically crucial to ask what means are justifiable in trying to reach that objective. At present, some 80 percent of the population of many African nations is involved in agriculture. While the percentage of farmers is lower in Asia and Latin America, it does not begin to approach the single-digit proportions of Europe and North America. Bankruptcy, abandonment of farming, and migration to urban areas will cause untold suffering among those affected, but there are additional moral costs that may be even more important.

A Few More Words on Behalf of the Smallholder

Sen's pathbreaking work in the theory of human development emphasizes capabilities over wealth. Capabilities are practically realizable actions or doings that people might undertake to improve their lives. A capability that is realized creates a "functioning" or a domain of individual or social life in which needs and desires that meaningfully improve quality of life are being actualized. Here we have been focused on food. A food entitlement is a capability; it indicates how a person gets access to food. But eating a steady, healthy, and satisfying diet is a functioning. It is a particularly important functioning that everyone will choose to realize whenever the capability for realization is secure.[19] Part of the reason that Sen makes a distinction between capability and functioning is that people with limited means will still want to choose which among their many other needs and desires to realize. Educational opportunity or even health care may be foregone willingly in some cases.

Along with Sen himself, David Crocker has emphasized the importance of *agency* among the capabilities that are the goal and focus of development. Here the word *agency* stands in for a fairly complex set of skills, conditions, and operational means that allow people to have some measure of control over the circumstances in which they live and work. Crocker is often most intently focused on political agency, the ability to express one's view and to have some impact on the processes of government and social decision making. Equally important is economic agency: the ability to undertake, direct, and influence the activities that secure the means for daily living. For many people in industrial society, economic agency depends on a robust job market, with numerous and diverse opportunities for employment. Industrial societies also secure a capability for entrepreneurship, a more radical form of economic agency where people take personal and financial risks with no particular guarantee of a corresponding reward. Relatively few may choose to convert their capability for entrepreneurship into a functioning in advanced industrial society, but the capability is nonetheless thought to be an important component of autonomy and personal freedom.[20]

While food security is one capability that is important for human development, agency is another. To undertake a pattern of growth that replaces smallholder production with industrial farming over the course of even two or three generations will annihilate much of the capability for agency that farming people currently have. Relative to wage workers who depend on someone else to provide opportunities for employment, farming represents a form of agency that is currently within reach for many of the world's poorest people.[21] To recognize agency as a capability that articulates the moral imperatives of development requires that we understand poor people as more than poor. It requires acknowledgment that even extremely poor people have skills that allow them to achieve a significant measure of self-help and self-reliance on a day-to-day basis. It means that we recognize them, in short, *as farmers*. A process of development that undercuts the capabilities of present-day rural people cannot be ethically justified by the long-term result, nor would it satisfy to come along after their lives and livelihoods have been thrown into turmoil by programs of agricultural intensification and then offer to secure their right to food by giving them a dollar or a loaf of bread.

Although many smallholders around the globe are quite poor, sometimes desperately poor, they may possess a great deal of economic agency. There is no one who tells them when it is time to go to work, and no one who tells them what to do. Although their farming choices may be constrained by their location, their technology, and their knowledge base, the crops they plant, the way they tend or harvest them, and the husbandry they practice is largely up to them. Even when small farmers are so poor that they lack a secure food entitlement, they nonetheless have a capability that will be weakened and even lost entirely when they exit farming to become wage workers. As unskilled workers in a minimum-wage economy, they are entirely dependent on someone else to create an employment opportunity. Lacking a job, they are dependent on the gifts and grants of government or charities to achieve their most basic functionings. In either case, they have lost the agency that they had as poor farmers to structure and control their daily work in pursuit of subsistence.

My claim here may be controversial, and it may also be easily misunderstood. There is no doubt that farmers in extreme poverty might gladly trade their situation for that of the low-wage factory worker. Such choices are made every day. The food insecurity of a farmer earning less than one euro per day may certainly be so great as to make wage employment at two euros per day highly attractive. What is more, there are cultures in which farming is viewed negatively, where farmers have little to no social prestige, and where leaving farming is viewed as a step up the social ladder. In such settings—not uncommon on the African continent—the smallholder's trial of poverty is exacerbated by a lack of dignity. It is no wonder that small farmers would eagerly accept wage-paying jobs in such circumstances. My claim is not that farmers *would* want to stay farmers, or that they *should* want to stay farmers. My claim is simply that *as* farmers (and even this is not true in every case) they possess a capability that will be weakened when they exit farming and enter the wage-earning class.

And yet there are also cultures where smallholding farmers are venerated. Farmers are admired for their resourcefulness and self-reliance. They are esteemed for their independence and for their diligence. Such virtues are tied to farming in Eastern and

Western cultures alike, and they are closely intertwined with the fact that farmers have economic agency. A society rich in farmers will, it is often thought, naturally attain a form of social independence and a recognition of the burdens that must be assumed in order to assure its continuance into the future. This is the thought behind Thomas Jefferson's famous praise of farmers:

> Corruption of morals in the mass of cultivators is a phaenomenon of which no age nor nation has furnished an example. It is the mark set on those, who not looking up to heaven, to their own soil and industry, as does the husbandman, for their subsistence, depend for it on the casualties and caprice of customers. Dependence begets subservience and venality, suffocates the germ of virtue, and prepares fit tools for the designs of ambition.[22]

It is widely known that Jefferson favored a republic of farmers. He feared a nation built on manufacturing. Manufacturers (or "artificers" as he called them) are "panderers of vice & the instruments by which the liberties of a country are generally overturned."[23] Jefferson may have held little regard for the capital-owning class, yet his concern was that those who depended on someone else to create a job for them would become both economically and politically dependent, would lack a sense of judgment, and would thus fail to perform the duties of the citizen. Jefferson's view may have been a bit overheated, but the point here is that he saw and valued the agency—both economic and political—that flowed from the farmers' means of making a living. The value of agency is thus both personal—as a capability enjoyed by many small farmers no matter how poor—and social—as a characteristic that makes the population more fit to perform the duties of democratic citizenship.

Jefferson's praise of farming takes us well beyond the fundamental problem in food ethics, yet small farmers' agency provides a supporting reason to view the industrial philosophy of agriculture skeptically. Solving the fundamental problem of food ethics by ushering farmers out of farming as quickly as possible may be quite shortsighted for two reasons. First, a poor farmer who leaves farming to participate in the market for unskilled labor gives up one capability in exchange for another. This trade-off may be attractive or even compelling to the individual who makes it. Giving up the capability for economic agency may seem like a small price to pay for the functioning associated with food security. But is this trade-off necessary? Would it have been possible for the individual to have remained a farmer (hence retaining a greater degree of economic agency) while also achieving a more secure food entitlement? If so, then that would be the superior outcome from a development ethics perspective, and it provides a reason to be cautious in endorsing strategies that move people out of farming.

The second reason—Jefferson's reason—presumes the hypothesis that a population with greater economic agency is more likely to exercise political agency in a way that achieves democratic functioning. Democracy is feared by those believe that people will vote for benefits but defeat the taxes needed to pay for them. Farmers are less prone to do this because their participation in a thick form of economic agency creates a natural appreciation of the relationship between finance and social sustainability. What is more, because farmers' assets are tied to the land, they see a relationship between the geographical boundaries of a polity and their own economic agency that does not apply to an entrepreneur with more portable assets. They must make democratic government work at a particular place. They can't go someplace else when taxes or labor costs rise. They have to make things work out wherever their farm happens to be. There is thus a social capability for democracy that corresponds to the proportion of farmers in the general population.

None of this is to say that poor farmers should be forced to farm against their will, and some of these arguments will be less persuasive where farming bears a cultural stigma. Nevertheless, these considerations provide a broader and more socially grounded ethic from which to evaluate responses to food insecurity that concentrate on the urban poor. There are reasons to help poor farmers be less poor but still farmers, and those reasons extend beyond the mere fact that farming is something they know how to do.

Furthermore, these reasons may be more important during the early decades of state building than after democratic institutions have taken hold. Political culture and tradition may suffice once democratic institutions have survived a generation or two, but societies that lack the experience with democratic political practice may sorely need the reinforcement of democratic political virtues that a farming people can provide. To the extent that a farming population contributes to either economic or political agency, there are even stronger reasons to take the fundamental problem in food ethics seriously.

And Finally . . .

The theory of development is a relatively new thing, though philosophical works such as Jean-Jacques Rousseau's *Discourse on Inequality* or Immanuel Kant's *Perpetual Peace* are forerunners. Many would give credit for introducing the idea of "development ethics" to Denis Goulet, whose book *The Cruel Choice* had just been published when Singer made his initial contribution to the ethics of hunger in 1972. Amartya Sen is not the only author to have significantly broadened and deepened our collective understanding of the ethical dimensions behind policies of economic and social development in the intervening decades. But as the idea of development ethics has taken root, the idea that hunger is a central or unique problem has waned. Food is viewed as just another good, rather like health care, education, recreation, and emotional attachment. Indeed, food is subsumed entirely into the domain of bodily health on Martha Nussbaum's celebrated list of capabilities—as if control over one's food entitlement is largely a medical problem.

Food producers have largely been erased from this picture of development ethics—shockingly so in light of the way that they dominate the global ranks of extreme poverty. The reasons for this are unclear, though the way that people in industrial societies are cut off from any concrete experience of food production must be partly responsible. The upshot is that the tension between food producers who are poor and face desperate challenges to their own food security (while simultaneously enjoying a surprising degree of agency and self-reliance) and the needs of the urban poor is largely unappreciated. The fundamental problem of food ethics is an ethical problem because it complicates any effort to act on behalf of the poor. It therefore tests any proposal to specify the duties of the developed world, or better-off people. But it is also an ethical problem that bleeds rapidly into exceedingly complex social and economic issues. It is not a problem that suggests easy answers.

We should be suspicious of grand ideas that resolve the tension between food producers and food consumers with a sweep of the pen. Although I am painfully aware that ethics will not *solve* the problems of food insecurity, I hope that I have done something in this chapter by showing that this tension abides as a fundamental moral problem. Philosophical arguments demanding that the well-off bring aid, overgeneralized interpretations of capabilities, and sweeping critiques of neoliberalism have not dissolved the fundamental problem of food ethics. Our first responsibility in a globalizing world is to see this *as* a problem that has not gone away with modernization.

Chapter 5

Livestock Welfare and the Ethics of Producing Meat

No topic in food ethics has attracted more attention from philosophers than the consumption of animal flesh. People have debated the ethics of eating meat since antiquity, but the nature of the debate has varied significantly over time. At the dawn of Western science, Europeans thought that vegetarianism was biologically impossible, but when their encounter with India revealed an entire nation *practicing* the impossible, a gradual re-examination of some once unquestionable cultural assumptions began. More than a few people who come to question the practice of taking an animal's life in order to feed on its flesh (like Henry David Thoreau) find themselves falling a little in self-respect. Many become vegetarians as a result.[1]

The ancient Greeks had debates about the similarities and differences between human beings and nonhuman species in terms that are strikingly familiar to contemporary ears. Aristotle held that humans alone were rational animals, while Stoics found much evidence for rational behavior in nonhuman animals.[2] Rational thought may not in itself imply the capability for moral discrimination, but even though few nonhumans qualify as moral agents—creatures that act on the basis of moral reasons—perhaps they nonetheless deserve moral consideration and respect. And wouldn't that also imply that they shouldn't be eaten? Although this inference may be a bit hasty, many philosophers of the ancient world practiced some form of vegetarianism. In some cases, a meat-free diet may have flowed from the type of dietary *ascesis* discussed in Chapter 3, but it is also clear that some ancients based their practice on consideration of respect for animals that many of their fellow citizens were quite happy to kill and eat.

The Return of Animal Ethics

The core ethical questions with respect to eating meat have been subjected to renewed reflection and debate during the last fifty years. Ruth Harrison's book *Animal Machines: The New Factory Farming Industry* was published in the United Kingdom in 1964. It precipitated the formation of a committee under the direction of embryologist F. W. Rogers Brambell to review the welfare of animals being raised in CAFOs. The Brambell Committee issued its report in December of 1965, stating that all food animals should have key needs met in any production setting. They were eventually formulated in terms of five freedoms:

1. **Freedom from hunger and thirst**—by ready access to fresh water and a diet to maintain full health and vigor
2. **Freedom from discomfort**—by providing an appropriate environment, including shelter and a comfortable resting area
3. **Freedom from pain, injury, or disease**—by prevention or rapid diagnosis and treatment
4. **Freedom to express normal behavior**—by providing sufficient space, proper facilities, and company of the animal's own kind
5. **Freedom from fear and distress**—by ensuring conditions and treatment that avoid mental suffering[3]

Harrison's book and the Brambell report stimulated a group at Oxford University to begin new scientific and philosophical discussions on the moral status of nonhuman animals. Some participants in this group published a collection of essays on the topic in 1972.[4] Peter Singer's review of this volume, published in the *New York Review of Books* in 1973 under the title "Animal

Liberation," brought widespread attention to the new era of philosophical reflection about nonhuman animals.

Applying an ethical rationale that bore important similarities to
the argument he had developed in connection with famine, Singer
argued that animals do feel pain and that whatever ethical theory
one is inclined to accept requires reasonable sacrifice when it can
avoid great evils. Hence, he concluded, it is time to re-evaluate a
host of human practices in respect to the nonhuman animal world.
Singer went on to argue that only unthinking prejudice could lead
someone to deny the first premise of this argument. His pattern
of argument suggested that the liberation of animals would follow
naturally on the heels of other liberation movements of the 1960s
and 1970s, such as those in opposition to prejudice against blacks
and other racial groups, and against women. The 1973 essay soon
morphed into the first chapter of Singer's influential book *Animal
Liberation*, which has been revised and republished many times
since its 1975 appearance.

Within a comparatively short time, Tom Regan advanced the
view that Singer had not gone far enough. Singer claimed that
humans and nonhumans are alike in terms of their capacity for
suffering, but Regan claimed that the similarity runs much deeper.
Like humans, many animals (certainly all vertebrates) are *subjects
of a life*. As discussed in Chapter 1, Regan's rights view implied that
as an individual being, every animal has the integrated type of
subjectivity that rights theorists have always thought to be the
source of moral significance. Regan held that all vertebrate animals are individuals who have a life history, and their mental life
is the basis for a personal identity that endures over time. They
have attachments and in some sense care about the future. It is not
sufficient to determine ethical obligations to nonhuman animals
simply by weighing their suffering against the benefits humans
get from using them. The ethical practice of thinking about a
nonhuman animal through the kind of cost-benefit trade-off lens
characterized by Singer's utilitarian approach is *already* a form of
disrespect for the integrity that each animal possesses as the subject of a life.[5]

Much of the philosophical debate that ensued centered on the question of moral status: do nonhumans deserve moral respect *at all*? Or, to put it slightly differently, are we obligated to take the interests of nonhuman animals into account when making decisions or taking moral account of our behavior? This question intersects with other philosophical questions about the nature of consciousness and mental life. If one believes that nonhumans lack a conscious mental life (if they are more like machines than people), then the question about nonhuman animals deserving moral respect could be answered easily in the negative. Philosophers from René Descartes (1596–1650) to Donald Davidson (1917–2003) have associated the peculiar nature of human consciousness with the ability to use language and have thus defended a radical separation between human beings and other creatures in the animal kingdom.

However, if some nonhuman animals *do* have a mental life (which seems beyond dispute to *this* author), the philosophical game is on. A review of philosophical positions on the moral status of animals since the 1970s provides one entry point into food ethics, and it is one that many recent teachers and writers have taken. Once one concedes that animals of the species humans use for food are sentient—that they register feelings of pain and well-being—it is impossible to deny that vegetarianism is a topic for food ethics. Meat eating and other food practices that entail breeding, slaughter, and keeping animals on farms for their entire lives are inextricably tied to harm. The legitimacy of eating animal flesh is a question that has sparked philosophical thinking since the time of Pythagoras (570–495 BCE) or Porphyry (234–305). Many people who consider the ethics of meat eating have concluded that farm animals suffer greatly from the use that humans make of them, and that the benefits cannot possibly justify the consumption of products that require the animals' deaths. The argument for vegetarianism may even be extended to other products of animal agriculture that do not require the death of the animal, such as milk, eggs, and wool. Although *these* products can be obtained without slaughter, production for milk, eggs, and wool are so integrated into meat production in the modern world that this difference is probably negligible. The list of philosophers advocating vegetarianism over the last decade only begins with Singer.

There are easily hundreds of published treatises on the subject, and they are overwhelmingly in favor of the vegetarian position. However, as Bernard Rollin has argued, whatever philosophers may have thought, livestock producers themselves have never doubted that the animals under their care are capable of experiencing pain, fear, and other forms of mental distress. There has never really been any serious question that the animals we use to produce food deserve moral consideration, at least among those who have some experience in animal agriculture.[6] Yet almost to a person, people who are professionally involved in producing food animals do *not* infer a duty to be vegetarians from their moral status, so it is empirically evident that conceding the basic points about animal consciousness hardly settles the question. Precisely because the arguments for and against ethical vegetarianism are so well represented in the existing work of philosophers, it is more important to devote this chapter to a series of questions that have rather surprisingly escaped a great deal of philosophical analysis.

The Ethics of Farm Animal Production

Food ethics could and should address at least three somewhat distinct questions with respect to livestock production:

1. Is it ethically acceptable to eat animal flesh, or to raise and slaughter livestock for animal food products?
2. Are present-day methods for raising livestock ethically acceptable?
3. How should present-day livestock production systems be reformed or modified in order to improve animal welfare?

There is an implied hierarchy in the way that many people (and certainly most philosophers) approach these three questions. People assume that question 1 is the most ethically important question, and given this assumption, one's answer to it shapes the way one approaches the other two. Question 1 is certainly the old one, debated by philosophers and other moralists since time immemorial. CAFOs, however, have origins in the holding pens that developed around industrial slaughter plants during the nineteenth century. Animals did not spend their entire lives in the confined spaces of a CAFO until the mid-twentieth century. One should not presume that animals were necessarily treated well in the past, but the point of the historical contextualization is simply to show that while question 1 is an old one, questions 2 and 3 presuppose the existence of industrial production systems. They could not have arisen until comparatively recently. So it is easy to see why there are no direct philosophical precedents for questions 2 and 3.

But there is more to the implied hierarchy in the way that these questions are addressed by most academic philosophers (and indeed by many ordinary people). First, suppose that your answer to question 1 is, "No, killing and confining animals for food production is *not* morally acceptable." If you have concluded that you shouldn't eat meat on moral grounds, the next practical step is to purchase a vegetarian cookbook; it is *not* to go on and consider the merits of question 2. If confining or killing animals is ethically unacceptable under *any* normal conditions (that is, ruling out the "stranded on a desert island" thought experiments), there is nothing additional to say in response to question 2. And if killing and confining animals is unacceptable under any circumstance, there is no way to fix an industrial production system, and hence nothing interesting to say about question 3. It is, in fact, only those who answer question 1 in the affirmative who move on to question 2. At this juncture, they have tended to follow the path laid out originally in Harrison's book and by the Brambell Committee, and they answer question 2, "No, industrial systems are *not* morally acceptable." Given the pattern established above, they never consider question 3. Instead of buying the vegetarian cookbook, they look for some alternative kind of animal production that *is* acceptable and they purchase meat from that source.

This is quite reasonable from the standpoint of a practical approach to one's personal diet, but it overlooks the fact that question 3 is logically independent from the other two. Virtually everyone to have considered animal use from a social perspective, including the most widely known advocates of ethical vegetarianism, concedes that animals will continue to be produced in CAFOs for some time. Pro-vegetarian social movements may catch on, but even the most ardent advocates admit that they won't catch on overnight. There may be some day when legislatures will ban industrial production, but that day is not on the immediate horizon, even in those European countries that have passed legislation targeted at certain livestock production practices. Thus, even if one thinks that *all* animal production is ethically unjustifiable, and even if one thinks that all CAFOs are ethically unjustifiable, it is still *quite* meaningful to ask how to improve the conditions for animals being raised in CAFOs. And crucially, this question is meaningful *from the animal's perspective.*

However one answers questions 1 and 2, there are still questions that will arise in trying to decide *how* present-day husbandry practices could and should be reformed. [7] Addressing these questions requires someone to have some kind of account of what it would take to improve the welfare of an animal in a given situation. In comparison to the philosophical work on question 1, there is very little work being done in the conceptual space implied by this question. One difficulty arises in connection with the fact that we are considering species that have considerably different physiology, neurology, and behavior than human beings. Question 3 raises "What is it like to be a bat?" questions. At about the same time that Singer was starting a new round of ethical reflection on moral duties to animals, Thomas Nagel posed this seemingly factual question. He argued that given a bat's reliance on echolocation as its dominant perceptual sense, it is very difficult for us to gauge a bat's subjective experience empathically. Yet we do not doubt that there is *something* that it is like to be a bat.[8] Asking what it would take to improve conditions in CAFOs requires us to consider an analogous question for pigs, cows, chickens, and other livestock species, but in a context where there is much more than a philosophical point at stake. Having any insight into question 3 requires that we ask a set of speculative questions about the subjective experience of farm animals, and we should not assume that their subjective experience is just like ours. Fortunately, scholars in neuroscience, animal behavior, and veterinary medicine have compiled a now considerable body of research that bears on this problem, but there are still gaps and philosophical challenges.

A second difficulty is that "improvement" presupposes change in existing systems, but question 3 attains its significance because we recognize socioeconomic constraints on this change. We would not ask how to improve conditions for livestock if it were possible to wave a wand and suddenly make everyone into vegetarians (as the writing of many pro-vegetarian philosophers seems to assume). Question 3 becomes meaningful precisely because it is practically and politically possible for *some* modifications in livestock facilities and husbandry methods to be implemented, even while the range of possible change stops somewhere short of everyone becoming vegetarian overnight. Notice that the transition from question 2 to question 3 also implies that the horizon for change stops short of everyone purchasing their animal food products from an ethically justifiable type of farm. If implementing immediate change in industrial animal production (and the buying habits of consumers who purchase meat, milk, and eggs) were trivially simple, there really *would* be no reason to consider anything beyond question 2. However, setting the appropriate parameters on the flexibility of the food system for the purpose of improving farm animal welfare is a task that requires us to synthesize empirical social science with philosophical judgment. The way that anyone undertakes such a synthesis will be open to debate, but *this* is a debate that is very rare in recent food ethics.[9]

Three Domains of Animal Welfare

The Brambell Committee's Five Freedoms was a framework for evaluating whether animal production is morally acceptable, but it is important to notice that in practice it also functions as a framework for considering how animal welfare can be improved.

Consider freedom from pain, injury, and disease. No animal, human beings included, lives a life that is *free* from pain, injury, and disease. While the language of "freedoms" connotes the total absence of the unwanted condition from the animal's life, this is not a reasonable interpretation of the Five Freedoms. What freedom from pain, injury, and disease means in the context of livestock production is that animals should not be kept under husbandry conditions where pain is a persistent element of the animal's experience, or where injury and disease are unrelieved by proper veterinary care.[10] Similarly, the freedom from fear and distress calls for "conditions and treatment that avoid mental suffering." This is often interpreted in terms of stress, but animals (again, including humans) experience short periods of physiological stress during peak moments, such as sexual orgasm. This is not a kind of stress that we want to eliminate from an animal's life. The freedom from fear and distress is thus not an absolute condition but a gradient that suggests relative levels of well-being. Even a straightforward criterion like "ready access to food and water" can be translated into an indicator for improving husbandry along a gradient. How many feet or inches should an animal need to travel between the drinking spout and the feed trough? The answer is not zero, for moisture would spoil the feed!

It is also important to notice that there are possible points of tension among the five freedoms. "Normal behavior" requires company of the animal's own kind, but other animals of one's own kind can also be a source of fear, distress, and mental suffering. Individual animals engage in pecking, biting, and butting in order to establish dominance. While it is typical for such behavior to subside once dominance is established, there are individuals in all species who engage in what we might call obsessive or compulsive persistence in dominance behavior. Indeed, the specification and evaluation of what is normal or typical in animal behavior (again, no exception for humans here) becomes a significant problem for animal ethics, and one that tends to be underappreciated by philosophers who are making the case for vegetarianism. This is a point that will be visited again below.

Collectively the five freedoms stipulate a framework in which relative states of well-being can be evaluated, but they do not imply absolute criteria that a livestock producer can meet fully and unambiguously in every instance. We might be tempted to think of the Five Freedoms as rights or entitlements that are analogous to human rights and entitlements, but we should recognize that overgeneralizing an indicator of well-being can be an ethical mistake. While some human beings in contemporary society endure mental suffering because they lack for company, others voluntarily suffer considerable mental suffering *as a result* of the company they keep. We don't have a clear way to say how the multiple criteria being articulated by the Five Freedoms could amount to an acceptable quality of life for ourselves and other human beings. We should not expect that it will produce something like that for other species, either.[11]

During the two decades following the Brambell Committee's recommendations, specialists in animal behavior and veterinary medicine conducted a debate over the criteria for animal welfare that was worthy of the most arcane and pedantic debates in twentieth-century analytic philosophy. Animal welfare is about *feelings*, some would claim; it is about the animal's conscious life. Others would counter by asking whether that means the animal has no welfare when it is asleep. As the twentieth century drew to a close, the scientists working in this field came to recognize that a farm animal's well-being was a complex blend of indicators, some identified by the Brambell Committee and others not made particularly obvious by the Five Freedoms approach. Ordinary elements of veterinary health are left rather implicit by the freedom from pain, injury, and disease, for example. In addition, animal welfare science came to acknowledge that there was an ineliminable role for ethics in combining and prioritizing these multiple elements of well-being.[12]

A "consensus approach" summarizes the diverse elements of an animal's well-being in terms of three broad categories (see figure 3). First, it is recognized that biological indicators of health are a major component of welfare. Animals of any species suffering from

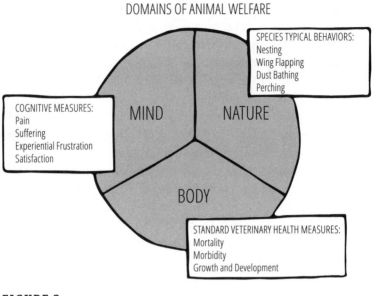

DOMAINS OF ANIMAL WELFARE

SPECIES TYPICAL BEHAVIORS:
Nesting
Wing Flapping
Dust Bathing
Perching

COGNITIVE MEASURES:
Pain
Suffering
Experiential Frustration
Satisfaction

MIND

NATURE

BODY

STANDARD VETERINARY HEALTH MEASURES:
Mortality
Morbidity
Growth and Development

FIGURE 3

Adapted from Michael C. Appleby, *What Shold We Do about Animal Welfare?* (Oxford, UK: Blackwell Science, 1999).

morbidity and mortality as a result of disease, injury, or the conditions in which they live have a compromised welfare. Other biological indicators of individual welfare include growth, respiration, and other types of biophysical functioning that can be normalized for the species. The calculation of these statistical norms can be tricky for farm animals because many of the animals used in agriculture have been bred to possess traits that are far from typical of their wild or "unimproved" conspecific relatives. This problem notwithstanding, the category of biomedical or veterinary health measures represents a relatively obvious domain of welfare for all animals. It is noncontroversial in that it is a domain that all livestock producers would recognize as valid.

Second, there are dimensions of welfare that derive from the way an animal feels. Affective states such as pain and pleasure or more complex emotionally charged experiences such as fear or sexual orgasm are almost certainly widespread across the animal world, and there is little reason to doubt that farm animals are capable of having such feelings. Measurement and characterization of these mental states is difficult enough for humans, but it is an exceedingly difficult task for animals with a substantially different genetic and neurological makeup. Nevertheless, as Nagel argued, we may find it difficult to imagine what it is like to be a bat, but we do not doubt that there is *something* that it is like to be a bat, some qualitative texture to every animal's experience.[13] We might characterize the domain of affective or experiential states as referring to animal minds, which suggests that the veterinary health indicators refer to animal bodies.[14]

The scientists who have developed this heuristic note a third category beyond those of animal bodies and animal minds. They observe that under some forms of husbandry, animals are unable to perform some of the behavior regarded as typical for the species. For example, the wild jungle fowl from which domestic chickens have been bred engage in frequent perching on sticks, branches, or rocks available in their natural habitat. Domestic chickens who have an opportunity to sit on a perch will also exhibit this behavior, but chickens who live in production environments where no perching places are to be had do not. Ever since the original Brambell Committee, research on animal welfare has recognized that the ability to perform these species-typical behaviors is a component of animal welfare. In work pioneered by David Fraser and other behavior experts, a third domain of welfare is specified to acknowledge that the ability to engage in such behavior is important. The scientist and animal activist Mike Appleby has referred to this category as animal natures.[15]

Although the five freedoms continue to be discussed, the bodies-minds-natures rubric has advantages as a tool for considering how the lives of farm animals might be improved. First, nothing from the five freedoms is eliminated. Freedom from hunger and thirst are just combined with freedom from disease and injury to make up the bodies category, while freedom from pain and distress are combined in the minds category. In fact, the emphasis on cognition or feelings arguably clarifies what is ethically intended by "distress." Second, the three domains more clearly indicate that there are positive dimensions to welfare, things that need to be done on an animal's behalf. This is perhaps most relevant in the domain of feelings, where the importance of providing for satisfying cognitive experiences may be a significant expansion beyond mere avoidance of pain, fear, or distress. Finally, the three domains provide a more useful framework for highlighting the ways in which dimensions of welfare may conflict with one another, helping us to think of welfare as a balancing act rather than just a matter of not restricting an animal's freedom.

Improving Farm Animal Welfare: Some Thought Experiments

Before moving on to a general discussion of the ethics of food animal production, it will be useful to consider how the bodies-minds-natures framework might be used to evaluate some specific animal husbandry practices. Some of the examples offered here are real-world cases. They have been implemented on a very large scale involving dozens or hundreds of livestock producers and millions of individual animals. Others are true thought experiments. In the latter case, the recommended changes have never been implemented, and the predicted impacts are thus speculative.

There are cases where the framework suggests unambiguous action. Prior to the year 2000, virtually all commercial laying hens in the United States were housed in cages allowing approximately 48 square inches (≈310 square centimeters) per hen. This housing system deployed a battery (or long row) of cages suspended in tiers so that automated egg collection as well as feed and water delivery could operate along the length of the line.[16] A typical cage housing four hens might have been about 14 inches (35.5 centimeters) in each dimension. This stocking density had been derived purely on the basis of profitability: it was the way for producers to get the largest economic return on their total investment, which included not only chickens and feed but also the extremely expensive machinery for collecting eggs, distributing feed, and disposing of manure.

Animal welfare scientists deploying the bodies-minds-natures framework conducted a series of studies which showed that unambiguous improvement could be achieved in indicators from all three domains if birds were stocked at a density of between 68 and 72 square inches (≈440 to 465 square centimeters).[17] This is hardly *ideal* for hen welfare. It provides just enough room for the birds to stand on the floor without crowding and turn around.[18] It does not provide enough space for hens to extend their wings or to engage in other species-typical behavior. The figure of 68 square inches was derived by emphasizing improvement in every dimension of hen welfare, thus minimizing the opportunity for disagreement among those who would privilege behavior (animal natures) or cognitive welfare (animal minds) over veterinary health (animal bodies). The lack of ambiguity was decisive in forums convened for voluntary reform of egg production in the United States. The McDonald's restaurant chain used this figure to specify production rules for its suppliers in the late 1990s, and the United Egg Producers (UEP)— the trade organization that represents commercial producers who supply grocery stores with eggs—followed suit in 1999.[19]

The example of egg-laying hens shows that the framework can be used to think through the prospects for improving animal welfare, and the UEP example shows that at least in some cases, producers are willing to implement the improvements. But not all problems are as unambiguous as recommended stocking density. For example, UEP has continued to utilize an animal welfare scientific advisory committee to make further reforms, including the abandonment of restricting feed in order to force hens into a molt. Hens will naturally stop laying eggs when the weather turns cold and will shed feathers (i.e., molt) in anticipation of spring, when they will grow new feathers and resume laying eggs. By design, hens in industrial systems lack seasonal signals for molting, but can be induced to molt if producers simulate the scarcity of feed that the onset of winter would bring. However, the hunger hens experience during feed restriction is clearly stressful from a cognitive standpoint (it is also contrary to the Five Freedoms), hence the practice of inducing a molt has been abandoned. Instead, the pace of laying declines gradually from a peak of roughly one egg per hen per day, and producers will "depopulate" the house (i.e., they will remove the hens and kill them) at the point when it is no longer profitable to feed them. If the birds were to endure a molt, they would return to a higher rate of lay and would subsequently have up to an extra year of life. There is thus a trade-off between cognitive well-being (the subjective experience of hunger) and the animal bodies goal of extending life.

If it is less clear that eliminating the feed-restricted molt improves the hens' welfare, there are other cases that are even more difficult. As already noted, the move to 68 square inches per bird could hardly be said to have offered hens the range of movement that they (or their evolutionary forbearers, the wild jungle fowl) would have had before industrial animal production was invented. So why not just get rid of the cages altogether? As most readers almost certainly know, it is entirely possible to produce eggs in "cage-free" or "free-range" conditions. One such system simply has all of the birds wandering around on the floor of the barn, while another resembles a battery cage system where the doors have been removed from the cages, allowing birds to lay eggs in the cage-like areas and fly down to the floor and take a stroll whenever it suits their fancy. The economic disadvantage of these systems is that it is harder to collect eggs. Some birds (though not many) are going to drop an egg in a place where the machinery won't find it. You can't just leave eggs lying around indefinitely if you are hoping to sell a product that meets modern standards for sanitation and safety; you must have workers moving through the barn looking for stray eggs. Cage-free systems are much more labor intensive, and the eggs that come out of them must be sold at a price that allows producers to recover those costs.

But perhaps those are costs that we should simply agree to bear? This response has won the day in Europe, where caged layer systems like the ones described above were banned by regulatory fiat. The question is whether this is an unambiguous victory for animal welfare. The reason it may not be is that pecking order is one component of species-typical chicken behavior (animal natures). Chickens establish dominance by pecking at one another until one bird decides that it has had enough and becomes submissive. Then the pecking will either stop or at least become less frequent. This is well and good in the flocks of ten to twenty birds, as might be observed among wild jungle fowl, and it is probably tolerable in a flock of forty to sixty birds that might have been seen on a typical farmstead in 1900. It is also no problem when the birds are in cages where the group size ranges between three and fifteen birds (some of the newer systems have very large cages that accommodate up to sixty birds). But a cage-free/free-range commercial egg barn will have between 150,000 to 500,000 hens occupying the same space. If you are a hen at the bottom end of the pecking order in an environment like that, you are going to get pecked. A lot.

The cage-free/free-range system may provide good welfare for the average bird, but it is almost certainly much worse than a cage for the least dominant individuals. It is possible that this is a somewhat optimistic estimate, however. In order to make the point, we will leave the realm of what is now thought to be known about chicken welfare and take an excursion into armchair thought experiments. The pecking order presupposes that hens can recognize one another and remember who is the boss of whom. Behavioral observation shows that they can almost certainly do this in a flock of sixty birds, but there may be cognitive limits that are reached when flock sizes trend upward. Even small, organic egg producers who supply co-ops, and farmers markets will have 1,000 to 5,000 hens. What is it like to be a chicken in a flock of 1,000? Of 5,000? Of 250,000? Speculatively, even the dominant birds endure a form of cognitive disturbance or agitation as they persistently try to establish their position among a crowd of this size. They may be undergoing a constant source of stress and anxiety simply by "ranging freely" in groups that far exceed their capacity for establishing a stable pecking order.[20]

Now, it is important to qualify this speculation by admitting that we (by "we," I mean humans) simply don't know. The philosophical point to notice is that there is almost certainly a trade-off between animal natures and animal minds. Establishing a pecking order is certainly a normal or natural thing for chickens to do. We are respecting animal natures when we opt for cage-free/free-range egg production. At the same time, we are definitely *not* respecting the less dominant hens' experiential welfare (the welfare measured in the animal minds category) with this type of facility. What is more, if the pecking reaches a sufficient level of violence, we are not respecting their bodies, either. Injuries and mortality from pecking is a problem even for those smaller-scale (1,000 birds) egg farms, and higher mortality in cage-free environments gets treated as a cost of business by UEP producers. If this speculative thought experiment has any validity, *all* of these birds would be better off in cages.[21]

This philosophical problem thus has practical implications for how we think about improving animal welfare. There are costs for getting it wrong, and the costs are borne mainly by animals. From a human standpoint, it is ironic that some of things that consumers do to ease their consciences about purchasing animal products may turn out to make things worse. From the hen's standpoint, it is not ironic. It is probably awful. The examples given here are only a few among many instances suggesting philosophical tensions that deserve far more attention from people who think of themselves as animal advocates.

Better Answers?

Some readers are almost certainly reacting to the conundrums just discussed by shaking their heads and saying to themselves: Doesn't all this simply prove why CAFOs are morally unacceptable? Doesn't this just show why, if not vegetarianism, then some kind of alternative and human livestock production approach is the answer? I want to offer two responses to this thought. The first is simply, *why not?* Nothing that that has been said in this chapter is intended as a defense of CAFOs. I have not been trying to convince readers to give up their backyard chickens or their cage-free/free-range eggs.

As I said at the very beginning of this book, I am not here to tell you what you should eat. So if *your* answer is either give up eating meat altogether or finding meat, milk, or eggs at a co-op or market that you feel better about, more power to you. This brings me directly to my second response: if changing your diet is the way that you are reacting to my thought experiments, it shows that we are asking different questions. If ethics is supposed to be a discipline for asking better questions, this is a very significant point.

The matter of how we could make things even just a little better for the animals being kept in CAFOs is quite independent from whether eating meat is morally defensible (question 1), or even whether CAFOs themselves are morally defensible (question 2). I began this inquiry in animal ethics by arguing that the improvement question remains meaningful even if the answers to questions 1 and 2 are negative. CAFOs are widespread in the contemporary food system, and more are being built in the developing world every day. If it is possible to make improvements for the animals that will live in these facilities, then the people who build and operate them should do so. More generally, many of us live and work in organizations or institutions (the family, the state) that perpetrate morally unacceptable and unjustifiable acts. The fact that one is living in a morally imperfect world does not relieve one of the responsibility to make whatever small improvements one can. In the context of food ethics, doing something for animals implies thinking about question 3.

But, you might ask, doesn't it also imply thinking about question 1? Someone might object with the following argument: improving the welfare of animals in CAFOs just makes it easier for people to avoid the most important ethical question, question 1. Is it ethical to eat animal products in the first place. If the answer to this is no, then moderating the harm done to animals in industrial production facilities does nothing to move humankind toward a genuinely ethical resolution to the problem of animal abuse. John McDowell has written the following in response to anyone who would seriously propose to make the lives of animals being kept in industrial farming operations better:

Suppose someone said the project of eliminating Europe's Jews would have been a lesser evil if its victims had been treated with the utmost consideration and kindness in all respects apart from being deprived of a life, which would of course have been done, in this fantasy, as humanely as possible. Such a judgment could be seriously advanced only in the somewhat crazy environment of academic philosophy. It distorts the way how things actually were matters.[22]

Indeed, it might be better for CAFOs to remain as horrific as possible, for that would motivate more people to give up the use of farm animal products altogether.

To be sure, I have studiously avoided the claim that improving the livability in CAFOs makes the slaughter of animals for meat "a lesser evil." Nevertheless, if *no* current production system for farm animals meets ethically defensible animal welfare goals, aren't the ethical vegans—people who eat *no* animal protein—right after all? Many ethicists have concluded that the human species just needs to get off the animal habit. In fact, I see no ethically based reason to argue with those who choose ethical veganism, whatever their reasons for making this choice. If veganism is the *only* ethically defensible position for the entire human species, it still might not be practical in the sense of universal veganism happening anytime soon. I've argued that this provides a reason to consider question 3. But perhaps McDowell is telling us that this kind of practicality is the same as admitting that resisting Nazi extermination of the Jews was not practical or actionable in Hitler's Germany. If so, then anything other than veganism or vegetarianism is morally unreasonable.

What would it mean to suggest that *everyone* should, as a matter of ethics, not only consider question 1 but also answer it in the negative? Consider one implication: according to the World Bank, the criterion for "extreme poverty" is an average income of one euro per day, which works out to be about $1.30. When you earn $2.60 per day, you leave "extreme poverty" and are then (by World Bank standards) simply poor. "Poverty" is more difficult to measure in such unilateral terms, but in the United States an individual making less than about $32.00 a day ($40.00 in Alaska) is considered to be poor. Roughly three billion people in the world are below the poverty line, and about half of them live in extreme poverty. According to research by household economists, when people move from extreme poverty to just being poor, the main thing that they spend their second euro per day on is animal protein. They buy a little meat to eat, or possibly some eggs or (less frequently) milk. I do not have the temerity to present myself as someone who could speak on behalf of someone who lives on $2.60 per day, so I am loath to even try to justify the dietary preference of people who do. Nevertheless, the claim that *everyone* should consider whether it is ethical to eat meat implies that when the poor act on these newly possible dietary preferences, they are very probably doing something that is morally wrong.

One might reject the force of this rebuttal by pointing out that people in dire circumstances can be excused from what would otherwise be a universal obligation. Even Tom Regan is willing to throw his dog off a lifeboat if it is necessary to save a human life.[23] But this response misses my point. Excuses apply in extenuating circumstances, but the logic of excuses implies that the action itself is still morally wrong. A poor person might be excused for stealing a loaf of bread. Theft might be excused when a poor person's situation takes a turn for the worse, but in the case at hand, their situation has taken a turn toward the better. Under modestly improved circumstances, the extremely poor add a little meat, milk, or eggs into their diet. My claim is that there is something curious with a moral system that reclassifies legally and traditionally sanctioned conduct of people at the utter margins of society as something that *needs to be excused*. I see this as something like an abductive reductio ad absurdum for the universal veganism position.

A more traditional argument against universal veganism might be made by adapting the approach to animal ethics that has been developed by Gary Varner. Drawing heavily on R. M. Hare, Varner notes that most ethical decision making is intuitive and unreflective. Critical moral thinking is called for when the intuitions of common morality fail us, or when we have some reason to think that our intuitions are leading us astray. A great deal of philosophical work on animals is done under the assumption that the second of these criteria is satisfied for nonvegans. But Varner recognizes that this may not be true for *all* people who include some meat in their diet. Most people throughout history lived in circumstances of personal risk and difficulty, and their moral intuitions—their sense of common morality—reflected the kind of challenges they faced on a day to day basis. It would be *unreasonable* to have expected them to engage in critical moral reasoning about including animal products in their diet.[24] Something similar is arguably true for many people in contemporary society. I would include not only poor people spending a bit of their second dollar per day on animal protein, but also families in industrial societies who live near the margin, where children may go hungry now and then. They already have a lot to think about, and asking them to entertain a radical shift in their diets could only occur under some very special circumstances. As the work on food deserts tells us, the local fast-food chicken or hamburger restaurant may be a bargain for them, especially when we consider how parents are already working multiple jobs and have little time to cook.[25] This implies, as Varner suggests, that while veganism might be thought of as exemplary conduct on moral grounds, it should not be thought of as morally obligatory.

Or to put the point still differently, there is an Inuit saying sometimes attributed to the shaman Aua that goes like this: *The great peril of our existence lies in the fact that our diet consists entirely of souls.*[26] Inuit peoples live above the Arctic Circle where a plant-based diet is a biological impossibility. They have evolved a culture around hunting. The second half of question 1 about raising and slaughtering livestock would be meaningless to a traditional Inuit. Expecting them to answer the first part (Is it ethically acceptable to eat animal flesh?) in the negative would be absurd, but this does not imply that the Inuit never pondered the philosophical and spiritual implications of their diet. Contemporary Inuits who *can* shop at supermarkets and eat tofu might now experience eating traditional game species as a ritual participation in their cultural traditions. Yet the aphorism just cited suggests that Inuit people *do* associate the consumption of animal souls with a form of existential peril. Indeed, *many* religious traditions call for expressions of religious gratitude and metaphysical dependence prior to eating. The reference to Thoreau at the beginning of this chapter highlights the sense in which eating a sentient creature can spark a particularly poignant experience of spiritual vulnerability. There is, in short, a fairly broad terrain of possible ethical stances to take with regard to eating animal protein. These debates will continue to be an important theme in food ethics, and we should not presume that all thinking people will inevitably answer question 1 in the negative.

I am sure that none of this will be persuasive to many of the animal advocates among philosophers who are arguing for ethical vegetarianism. If one concludes (with Tom Regan) that all vertebrate animals are very much like human beings in terms of their subjective life *and* that this kind of subjectivity is the basis on which ethical claims are justified, there will be very little room for any ethical justification of meat consumption. The bodies-minds-natures framework is a start toward asking whether farm animals *are* very much like human beings, however, and it shows that we can err as much in assuming that they are as in assuming that they are not. In addition to the points already considered in our thought experiments, it is important to consider whether nonhumans have a mental life characterized by attention—an ability to narrowly focus one's consciousness that may have evolved in predators in order to facilitate tracking prey. It will also be important to ask whether they have episodic or merely semantic memory systems. The former may be necessary for the imaginative thinking we associate with planning for the future, but there is evidence that this capacity does not even develop among humans until late childhood. Animals lacking these capabilities may be quite capable of being harmed, but they may be very *unlike* "subjects of a life."[27]

More pertinent to keeping animals in farm settings, it will be helpful to have some sense of whether being able to socialize with your conspecifics is more important than being sure they will not get your food. One of the curious things about CAFOs for swine production is that most of the individual sows kept in isolated furrowing and gestation stalls do not exhibit any obvious perturbation,[28] yet humans would find this kind of confinement intolerable.[29] This is not an ethical justification of swine CAFOs. My interaction with specialists in pig behavior and husbandry leads me to think that we could almost certainly do much, much better. Yet my interaction with nonspecialists—including many philosophers—leads me to suspect that they believe all pigs in these stall facilities are enduring constant suffering tantamount to waterboarding or some other form of torture. Although we cannot be sure what it is like to be a pig in a gestation stall, getting better at articulating and unpacking what we suspect is actually going on in the lives of farm animals—and why it matters ethically—will be the next stage in food ethics.

The Ethics of Constrained Choice

McDowell's claim that working to improve the lives of farmed animals ignores how what is actually happening really matters might also be interpreted as a version of question 2: are present-day methods of livestock production morally acceptable? All of the alternative approaches to producing eggs that were discussed above actually exist at a commercial scale in the current socioeconomic environment. They are, at present, economically viable and represent meaningful options for commercial production. But would it be possible to implement a response to some of the problems discussed above by, for example, limiting flock size or returning to something that more closely resembles the kind of life that farm animals had in Europe or North America around 1900? What limits of practicality must we accept in order to ask the most meaningful questions in food animal ethics?

It may be too obvious, but it cannot hurt to begin with the observation that commercial livestock farmers are in it for the money. They may have some fondness for the animals they keep. The economics of beef production, in particular, are skewed by the many thousands of people who keep herds on pasture without paying too close attention to the bottom line. Nevertheless, when livestock producers say they are doing something "for the benefit of the animal," this does not mean that they would undertake the kind of effort or financial outlay that people expend for their pets. Livestock producers derive the money that they spend on raising their animals by selling live animals or animal products. At the end of the day, the money they take in has to exceed the money they pay out, for producers are also using their income to heat their homes, buy their clothes, pay their doctor bills, and send their children to college.

Whatever they produce, be it beef, pork, or organic beets, farmers in industrial societies must be able to recover the costs of production, or else they do not remain producers for very long.[30] This situation is sometimes described in terms of the need for profit, but it is more accurate to emphasize cost recovery. True profit occurs only after one has achieved a return on one's own labor that matches the market rate for comparable work,[31] plus a return on one's capital that exceeds the going rate of interest. A surprising number of agricultural producers are not profitable by this standard, though many of them will continue to stay in farming as long as they can pay their bills and eke out a living for their families. At the same time, industrial scale farmers, whether in crops or livestock, are not poor. Most achieve middle-class living standards, while a few achieve real wealth.[32] If it is possible to squeeze out a little higher return on one's livestock production operation by sacrificing the welfare of animals, the logic of the market dictates the result. When a few producers start to cut corners at the expense of their animals' well-being, the commodity market for beef, pork, broilers, or milk will reflect that in the price *all* farmers receive. Those who fail to squeeze out a higher return on their labor and capital will find that they cannot recover their costs, and soon everyone in the industry is using the same exploitative production methods.[33]

If animal welfare is seen as the ethical responsibility of the individual producer, there is really no way out of the cost-price squeeze. Farmers have economic incentives to maintain a minimum level of welfare for livestock because unhealthy animals do not bring the best prices. But in virtually every animal industry, there are numerous ways to increase one's return by doing things that decrease farm animal welfare. Farmers have no choice about this. If they choose to go against the standard industry practice in order to provide higher welfare, they are eventually going to go broke. They are effectively choosing not to be farmers. One of the most critical ethical lessons is that responsibility for improving animal welfare has to be distributed across the industry as a whole. Farmers have to agree on "rules of the game" that prevent everyone from sacrificing animal welfare, and they have to be assured that everyone actually plays by those rules.

There is no doubt that some trade-offs among the various dimensions of welfare are exacerbated by the scale at which industrial animal production now operates. Groups of forty chickens or fifteen hogs that might have existed on a diversified family farm of the early twentieth century afford more straightforward solutions to the welfare problems of dominance behavior, for example. But we must fix our reference point. The fact that these farms were diversified means that a considerable number of different production activities were competing for the farmer's time and presence of mind. Multiple species of food animals (cows, pigs, chickens, goats, sheep) were vying for attention with a field of cereal or fiber crops and a several-acre garden plot (with dozens of vegetables), not to mention some fruit and nut trees. Although *ideal* husbandry on small farms can potentially resolve many animal welfare issues, it is unlikely that animals on an average family farm of yore got ideal husbandry. As such, it is far from clear that they had significantly better welfare than animals on the average industrial animal farm of today. They suffered from untreated diseases, were left too long in the sun or without water, and were victimized by each other when not fending off wolves, hawks, coyotes, and snakes.[34]

The fact that contemporary animal producers have some sense of historical continuity with and personal memory of this past partially explains why they balk at broad social or governmental initiatives intended to improve animal welfare. It is also true that many farmers hate regulations and don't trust government. Even if farmers might be philosophically *willing* to support a change in the rules that improves the lot of animals, they have been dispositionally opposed to state action that would accomplish this. In Europe especially, animal protection groups have been successful in getting national and in some cases European Union-wide legislation passed to restrict certain types of farming operations on animal welfare grounds. They have been opposed by farmers, in part because farmers everywhere seem to distrust government but also because the rules have tended to be too inflexible, enshrining a particular doctrine about how to balance the various indicators of welfare and being unresponsive to changes in technology or husbandry. In the United States, the best achievement so far has been "guidelines" promulgated by the various commodity organizations—the National Cattlemen's Beef Association, National Pork Producers Council, Chicken Council, and National Milk Producers Federation. The UEP is far and away the most active on animal welfare at this writing.[35]

In the United States, change has come through voluntary standards coupled with actions by the retail sector that provide producers with incentives for compliance. The best known were initiated by McDonald's, the hamburger giant. McDonald's promulgated animal welfare standards for egg and pork production and insisted that their suppliers prove that they met these standards in order to be eligible for supply contracts. The model has spread across the chain restaurant industry and has been taken up by other major retailers such as Walmart. While it would be naïve to think that this approach can solve all the animal welfare problems that currently exist, it has some advantages over the regulatory approach that has been followed in Europe. Flexibility is the first. Standards can be adjusted as new knowledge becomes available. The fact that producers cooperate also means that the phase-in time for these new standards can often be very rapid. When the UEP decided to increase the space allotment for their hens in 1999, the new standard was being followed by 80 percent of the industry within three years. The regulatory standards banning cages in Europe had a thirteen-year phase-in period, and anecdotal reports indicate that less 20 percent of European producers may be in compliance even now that the phase-in period has been completed.[36] From the standpoint of the effect on animals' lives, it is at least debatable which of these approaches was more ethically justified.

There are other ways to relax the constraints that currently limit the potential for improving animal lives. It may be possible to breed a strain of chickens that are less aggressive, for example. The chicken breeds used in meat production are notably less troubled by the pecking order problem, though they have also been criticized by animal welfare advocates for being dull-witted, torpid, and generally un-chicken-like. They are also notoriously poor egg layers, which explains why they are not already used for egg production. Chicken breeders continue to search for a bird the combines a mild disposition with prodigious egg production.[37] In fact, breeding away animal welfare problems (or more extreme technological interventions like genetic engineering) fails to escape the ethical tension between animal natures, on the one hand, and animal bodies or minds, on the other. Animal breeding or genetic modification prioritizes the pain or anguish noted in the animal minds category and alleviates the pain and anguish by modifying animal natures in a radical way—by creating individuals that utterly *lack* a species-typical capability for pain or anguish. While this may, at the end of the day, be a defensible and even laudatory response, it is not free of ethical controversy.[38]

Ethical Consumerism for Meat, Milk, and Eggs

McDowell's claim that improving the lives of farm animals makes sense only in the crazy world of academic philosophy is in fact typical of that world, a world in which people routinely pontificate about nonhuman species with little knowledge of the conditions they actually live in, and even less understanding of what consequences would ensue if their proposals were put in place. In contrast, I conclude that there is a collective obligation for industrial societies to take steps toward reducing the animal welfare harms associated with industrial animal production.[39] This chapter has reviewed some of the puzzles that need to be addressed in taking such steps. Reducing harms would also require reversing a trajectory of growth in the production of animal products at the same time that one seeks to mitigate both environmental and animal welfare impacts of current methods. This would place the relevant ethical imperative in a domain with many other norms in environmental ethics, such as the obligation to reduce emissions of greenhouse gases in order to forestall (or even reverse) the impact of climate change.

Beyond this, it is certainly reasonable for anyone to take on a vegan or vegetarian diet. Further discussion of this would extend a chapter that has, perhaps, already gone too far beyond an introductory treatment.[40] There are many examples of arguments for and against vegetarianism circulating in present-day philosophical circles, and readers interested in these debates have many sources available. Yet cows, goats, pigs, sheep, chickens, and other poultry will continue to be raised for food for the foreseeable future. In considering how humanity might do better for those animals that will be kept for human food, we broach a number of considerations that complicate the assumption that old-fashioned farming was free of the quandaries we associate with an industrial food system. In light of these considerations, ethical vegetarianism starts to look like a retreat from a truly engaged and reflective evaluation of food. In closing this topic, I will reiterate a point made in passing above.

Too many arguments for ethical vegetarianism that circulate today involve simplifications and generalizations that neglect the multiple connections of foodways. Sometimes this neglect rises to the level of disrespect. The claim that eating meat is merely a "trivial pleasure"—something that could be readily sacrificed for the sake of nonhuman animal interests—is a case in point. It is profoundly disrespectful to tell someone earning $1.30 or less per day that the bits of animal fat or broth in their stew constitute a trivial pleasure, even if it is true that they could live without them. In fact, the view that what human beings eat is predominantly or even frequently a question of pleasure-seeking assumes a posture that distorts food ethics badly. Some people get very little pleasure from their food, and this is true even of wealthy people in industrialized countries. As the diet-health questions reviewed in Chapter 3 show us, many people find it difficult or even impossible to bring their food choices under control of the reflective mind. People who have become vegetarian after years of eating meat have been able to reinvent themselves and reform a set of behaviors that many other people could not even begin to contemplate changing. Although they may not have experienced this change as "easy," the fact that these individuals have done it implies that people who choose to be vegetarians late in life are among what may be a minority of the human species who can regard their food from a purely discretionary point of view. Advocates of ethical vegetarianism tend to write as if anyone could do this, as if it were as simple as changing shoes. Attendance at the intersections of food ethics suggests otherwise.

Chapter 6

The Allure of the Local

Food Systems and Environmental Ethics

In a recent book, Richard Hobbs, Eric Higgs, and Carol Hall define a *novel ecosystem* as

> a physical system of abiotic and biotic components (and their interactions) that, by virtue of human influence, differ from those that have prevailed historically, having a tendency to self-organize and retain its novelty without future human involvement.[1]

Hobbs, Higgs, and Hall intend to discuss a range of scientific and philosophical questions for conservation ecology in an era when anthropogenic climate change has rendered traditional norms and methods impractical. Preserving or restoring an ecosystem to its historically natural state becomes pretty difficult when it becomes too hot or too dry for the plants that used to grow there to survive. Nonetheless, it is striking how closely their definition of a novel ecosystem aligns with a term of art that is used in my corner of environmental ethics. Those of us who work in food ethics also study ecosystems with components that reflect profound human influence. We call them "farms."

To be sure, it is improbable that a farm would retain its characteristic interaction of abiotic and biotic components without continuous human involvement, but there are very few putatively natural ecosystems whose composition of species is independent of human involvement. Native American use of fire, for example, is now known to have had a profound impact on the "pristine wilderness" discovered by European settlers of the North American continent. Controlled burns in grasslands and forests alike opened smaller areas for cultivation and encouraged the growth of plants preferred by game species such as deer and buffalo. They also cleared the dense brush that becomes tinder for truly catastrophic fires, such as those in Western North America are now experiencing in a post-Smokey the Bear era of nature management.[2] Whether Native Americans who used controlled burns were farming depends on how one defines the word *farm*, but they were certainly using fire for the purpose of food and fiber production. I do not mean to argue against applying the concept of a novel ecosystem in conservation biology. Rather, I intend to highlight the way that human influence on ecosystem processes continues to trouble the assumptions of environmentally committed individuals.

For example, Laura Westra argues that agricultural areas should be regarded as buffer zones that protect natural areas from the impact of industrial civilization. She does not accept the idea that farms have any intrinsic value as ecosystems.[3] I once stood in a Mexican field of blooming cempasúchil flowers with Westra, who remarked, "Isn't it wonderful to be out in wild nature." The cempasúchil is a large and spectacular variety of marigold that is cultivated by Mexican farmers for decorative use during *Día de los Muertos* festivals. The field we were standing in was (to me) a farm: the flowers were all the same height and were blooming at the same time. It was, in fact, a monoculture of cempasúchil: there were no other plants in the field. In Laura's defense, it was indeed nice to be out there, and it looked nothing like a southern Ontario field of corn or soybeans. Whether farms are part of the natural environment or not turns out to be strongly influenced by culture and by aesthetic sensibilities. In Great Britain, a centuries-old tradition of public footpaths means that a day in nature will very often mean traversing a number of farms, while the blend of private property rights and legal liability tends to push North Americans into parks and nature preserves. It is possible to think that environmental ethics requires protecting ecosystems *from* agriculture, but it is also possible to valorize certain configurations of flora and fauna that have been profoundly affected by farming or grazing and to regard these configurations as the nature that needs to be preserved.

The lack of any clear line separating agriculture from the natural environment makes it difficult to stipulate or defend a universally applicable standard for evaluating the environmental impact of food production. Rachel Carson's *Silent Spring* documented the effect of using DDT to control agricultural pests on songbirds.[4] The 1962 book caused uproar over the unintended consequences of chemical use and is sometimes credited with sparking the social movement for environmental protection in the United States. Although pesticides have been regulated in the United States since 1910, the original intent was to protect farmers and other consumers from fraud. Carson's work was an important catalyst for reforms that occurred during the Richard M. Nixon administration. This led to the formation of the Environmental Protection Agency (EPA) in 1972 and to revisions of US regulatory policy that recognized the legitimacy of limiting the impact of pesticides on non-target species. This was, in effect, a policy that recognized the idea that agriculture could have untoward and unwanted impact on plants and animals within and surrounding a farmer's field. The new policies continued to recognize the legitimacy of killing fungi, insects, birds, or rodents when they were damaging the crop, but they also implied that collateral harm to non-pest species could only be justified by compelling benefits to producers.

These legal changes in the United States were occurring contemporaneously with actions that protected wilderness areas (1964) and endangered species (1966 and 1973). Although these policy actions were less clearly addressed to food production than to pesticide regulation, they reflected a shift in consciousness with worldwide implications. In 1987, the World Commission on Environment and Development (WCED) recognized the need to confront environmental challenges while promoting global development of natural resources that would meet the needs of all people, including future generations.[5] The work of the WCED was followed by the Earth Summit in Rio de Janeiro in 1992. The Convention on Biological Diversity (CBD) was opened for signature in Rio and now constitutes an internationally binding treaty that has had important impact on agriculture. The CBD recognizes that the global diversity of plant and animal species is a public resource that can be adversely affected by human activity. In one sense it is an extension of the philosophical rationale behind the 1973 US Endangered Species Act, which recognizes the loss of species diversity as a form of harm. However, the CBD's approach to biodiversity goes considerably further in seeking the preservation of all species, including those unknown to science, and in recognizing the value of genetic diversity—the existence of multiple alleles within the gene pool of any wild or cultivated species.[6]

The upshot has been the creation of entire subfields within economics and philosophy debating how species, ecosystems, and biodiversity are to be valued. The issues have implications for food ethics in at least two ways. First, the protection of biodiversity has been a key topic in the international debate over genetically engineered plants. Biotechnology is taken up in connection with Green Revolution approaches to food-system development in Chapter 7, while the relationship between biotechnology and sustainable agriculture is discussed in Chapter 8. Second, the growth of an environmental consciousness and a concern for sustainability lies at the heart of dissatisfaction with industrial agricultural production methods and the growing support for organic, local, and other types of sustainable agriculture. Many would argue that environmental leadership passed from the United States to Europe especially in respect to this latter development. European retailers, consumers, and governments have embraced "green labeling" and environmentally oriented supply-chain management. As discussed in Chapter 1, "food ethics" is equated with making food purchases that support environmental sustainability. The task of this chapter is to sketch a few of the ethical rationales that support such developments.

Agricultural Sustainability

More than thirty years ago and well before the WCED report sparked an international debate on sustainable development, Gordon K. Douglass assembled a group of people working in agriculture and food systems to consider the concept of sustainable agriculture. Their individual musings were published in the 1983 book *Agricultural Sustainability in a Changing World Order*. Reflecting on their contributions for the introduction to the book, Douglass noticed that his authors seemed to have three different things in mind. A number of authors were focused on *food sufficiency*. For them, the problem was all about producing enough food to sustain a growing global population. A second group of authors defined sustainability in terms of *ecological integrity*. For this group, the problem was that industrial production methods were overtaxing the ecological processes that have historically made it possible to understand food as a renewable resource. These production methods were consuming large amounts of fossil fuel energy, were threatening processes that replenish water supplies, and were vulnerable to catastrophic collapse from insects and from plant and animal diseases. Finally, there was a group interested in what Douglass called *social sustainability*. They were inspired by the work of anthropologist Walter Goldschmidt, who had studied two California towns in the 1940s. The town that was surrounded by small farms was thriving, while another located in a region dominated by large and economically successful agribusiness farms was struggling to retain its schools, social services, and local businesses. The "Goldschmidt thesis" was that large-scale farming was not conducive to healthy and vibrant communities.[7] A few years later, Miguel Altieri published the first edition of his book *Agroecology* in which he cited Douglass and produced a Venn diagram with circles labeled "economic," "ecological," and "social" (see figure 4).[8]

The discussion of sustainable agriculture predates the international debate over sustainable development, and it is worth noticing how it both intersects with that debate and how it develops different themes. It is entirely reasonable to assume that future generations will want to eat. Meeting the needs of the present without compromising the ability of future generations to meet their needs implies the need to anticipate global population growth and to gear present-day investments in agricultural technology toward producing enough food to feed more people. The driver is simple: the world needs or will soon need more food. The options for producing more food have historically involved bringing more land into

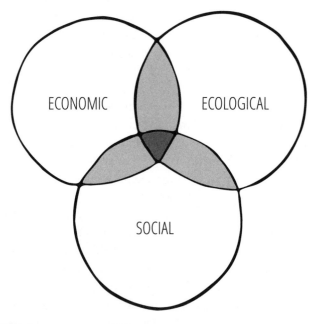

FIGURE 4

Adapted from Miguel Altieri, *Agroecology: The Scientific Basis of Sustainable Agriculture* (Boulder, CO: Westview Press, 1987).

agricultural production or figuring out how to increase the yield of existing acres. Expanding the area under cultivation runs smack into the environmental movement's commitment to preserving natural areas for purposes of recreation, biodiversity conservation, and intrinsic value. So advocates of the food sufficiency view generally emphasize the second option. Advocates of food sufficiency also tend to be agricultural scientists or capitalists who will benefit personally from developing new technology. While this provides a basis for being skeptical of their perspective, it should not obscure the fact that there *is* a compelling ethical argument for taking steps to anticipate future scarcity in the global food supply.

Supporting the needs of a growing global population will require manufacturing, mining, and other activities in addition to food production. All of these activities require energy and water, so agriculture will come into increasing competition for the use of these resources. (Recall that modern agriculture uses copious amounts of energy to produce synthetic fertilizers as well as powering machinery.) The food sufficiency perspective thus implies that agricultural production will have to become much more efficient: farmers and food producers will have to extract even more output from a dwindling supply of land, water, and usable energy. The only resource in abundance is sunlight. Indeed, the approach that many agricultural scientists and population experts take sees sustainability as a large and complex accounting problem focused on resource sufficiency.[9] The food and agriculture piece of this problem gets addressed by developing new methods of agricultural production that extract higher yields from the existing land base, without also increasing the input of water and energy or causing harmful impact on the supply of other human needs such as health and environmental quality.

In contrast to the food sufficiency advocates' emphasis on population growth and food needs, the advocates of ecological integrity and social sustainability were calling attention to problems that, in their view, had already been realized in response to the introduction of production-oriented technology in agriculture. On the environmental side, Rachel Carson's documentation of

the unintended consequences of chemical pesticides was exhibit
A. More broadly, the introduction of monoculture production systems conform closely to the ethical problem of "novel ecosystems"
mentioned at the outset of this chapter. If one values the configuration of biotic and abiotic elements in a natural ecosystem,
one can hardly imagine anything worse than plowing the whole
thing up and replacing it with orderly rows of corn and soybeans.
Agricultural insiders were also aware of problems associated with
the loss of soil through erosion and through harm to the microbial
ecology crucial to soil fertility. All of these environmental impacts
and more were being traced to the putatively more productive and
efficient methods of industrial agriculture. Advocates of ecological
integrity were of the view that these impacts were not a *necessary*
consequence of agriculture. They believed that alternative (though
perhaps slightly less productive) farming methods could sustain
agricultural ecosystems indefinitely.

The advocates of social sustainability were noticing a different
kind of impact from industrial monoculture: the destruction of
healthy rural communities. Like the ecologically minded critics of
industrial technology, they believed that these consequences could
be avoided by emphasizing norms and social policies that counter
some of the economic forces unleashed by unfettered capitalism.
The advocates for social sustainability also tended to think that
smaller-scale farmers would be more attentive to the environmental impacts that were of concern to the ecologists. There is thus a
sense in which both groups of critics—the ecologists and the social
activists—were presuming that a sustainable agriculture would
be one with farming methods and a social organization that could
be reproduced year after year, decade after decade, and generation
after generation. This vision did not imply the total lack of change,
but it did imagine both ecological and social systems as having
mutually supportive and regenerative elements that confer stability and resilience on good farming practices, as well as the overall
organizational structure of farming communities. Thus, both ecologists and social activists were equating sustainability with a kind
of functional integrity.

Resource sufficiency and functional integrity are not wholly distinct. Anyone who is interested in increasing global food production will be concerned about soil erosion and fertility. They will eventually get around to the ecology of these processes and when they do, they will be using the same scientific models as ecologists who foreground ecological integrity. Advocates of resource sufficiency may be less concerned about the decline of rural communities, but they certainly recognize that farming needs to be profitable and that it requires certain support services. The areas of contrast and overlap between resource sufficiency and functional integrity provide an interpretive framework for making sense of ethical debates on the environmental goals and impacts of food production. These two ways of looking at sustainable agriculture may differ most dramatically in terms of the way that they conceptualize the key values that arise in connection with agriculture and the food system. Developing this framework will require a few tangential discussions, however.

Sustainable Development: A Philosophical Interlude

As noted already, the 1987 report of the WCED had a singular influence on the way that environmental issues were conceptualized and debated. The WCED had been chaired by Gro Harlem Brundtland, former prime minister of Norway, and its report contained the most frequently cited definition of sustainable development: "Sustainable development is development that meets the needs of the present without compromising the ability of future generations to meet their own needs."[10] The Brundtland Commission (a common nickname for the WECD) had been convened as an independent body to develop a framework for thinking about environmental constraints on economic growth within the context of international agreements and development policy. Its influence has increased as we have come to understand how industrial growth not only depletes natural resources (such as petroleum and other minerals) but also causes pollution that will remain in the atmosphere for many decades. Polluting emissions may be toxic, causing asthma, cancer, and other degenerative disease. In the case of greenhouse gases, emissions may affect temperature, rainfall patterns, and habitats for beneficial insects (such as bees) as well as insect pests. The Fifth Assessment Report of the Intergovernmental Panel on Climate Change (IPCC) reports that global agriculture has already been affected by climate change. It notes that as sea levels begin to rise, some of the world's most productive cropland will be submerged.[11]

The emissions that are the source of these changes come from manufacturing, transportation, and household consumption of electricity and heating fuel. Some of them come from agricultural production itself. From an ethical perspective, greenhouse gas emissions are similar to the toxic effects of pesticide. There is a trade-off between the beneficial effects of food production and economic growth and the harmful effects of pesticides and carbon emissions. There is thus a profound sense in which the consumption of current generations is threatening the ability of future generations to "meet their own needs.". At the same time, as Chapter 4 discussed, people in less-developed parts of the world do not enjoy many of the comforts or forms of consumption that are commonplace in the industrialized countries. There is thus every reason to worry that even if industrialized countries pare back their polluting emissions—something that is proving to be politically very unpopular—industrial growth in the less developed parts of the world will cause total emissions to grow unchecked for the foreseeable future. While the expected environmental impact of past emissions is bad enough, the unchecked growth in pollution is potentially catastrophic. Yet would it be fair to stifle economic and industrial growth that improves the quality of life for people living in underdeveloped countries, especially while people in the industrial world continue to enjoy unsustainable levels of consumption?

The Brundtland Commission's definition of sustainable development has thus had a significant impact not only on international politics but also on the way that key questions of fairness are understood, especially as they relate to the rights and interests of future generations. Simon Dresner is typical of many scholars who have interpreted the debate over implementing the WCED definition of sustainable development as a debate over sustainability itself. His book *The Principles of Sustainability* follows this debate through a series of stages in which economists have debated whether achieving sustainability requires setting fixed limits on the consumption of resources, or whether changes in technology and knowledge will allow future generations to enjoy a better quality of life despite the consumption of present generations. He notes that this debate sometimes centers on how we should understand the idea of development. For example, some criticize the practice of measuring development in terms of growth in gross domestic product (GDP), a number that may not reflect *any* improvement in the quality of life for the poor. Others want to stress how *economic growth*—typically measured by GDP—does not necessarily imply problematic forms of pollution or depletion of natural resources, even if they have been connected in the past.[12]

These questions are important and they are also complex. Elsewhere I have argued that the debates over sustainable development tend to take place entirely within the framework of resource sufficiency.[13] The basic idea here is that a process or practice is sustainable if the resources needed to carry it out are foreseeably available. The thinking behind the Brundtland report was thus very much in line with that of Douglass's idea of food sufficiency. The practice that was the focus of the food sufficiency viewpoint was crop production, while the process that was the subject of *Our Common Future* was economic development. Shifting from crop production to a more abstract process like economic development has a profound effect on the debate, to be sure. Nonetheless, the underlying ethical assumptions about what matters and why are remarkably similar. Looking ahead to the rest of this chapter, the larger point is that some people think that *every* ethical question associated with sustainability and environmental ethics is subsumed by the resource sufficiency framework. If your thinking is completely dominated by worries about having enough (or conserving enough) in an era of climate change, you are very likely to think that the debates that followed the 1997 publication of the Brundtland report represent all of the important perspectives one could take on the environmental ethics of food. While I contest the totalizing aspects of this perspective, I certainly would *not* contest the claim that these are absolutely crucial questions to understand and address.

The Ethics of Resource Sufficiency

The key ethical issue that arises in connection with resource sufficiency is the classic problem of distributive justice: sufficient for whom? The industrial revolution has unleashed a sequence of powerful tools for increasing the efficiency of production processes, but the benefits of these innovations have seldom been shared equally. The Brundtland Commission was only one of many international committees that have been convened in response to the global inequality created by two centuries of colonialism. The economies of the global South were organized to channel raw materials and cheap food to Europe and then to successful industrial economies in North America and Australia. For the South it resulted in exploitation of natural resources, destruction of local ecosystems, the creation of oppressive political regimes, and the impoverishment of subject peoples. More subtly, this exploitation was accompanied by the creation of racial, cultural, and gender stereotypes that kept subject peoples servile and maintained a flow of goods to European power centers. A large share of recent social and political debate concerns the justifiability of these inequalities, as well as the appropriate contemporary response to it. Some authors argue that perpetuation of these injustices in the distribution of benefits is unsustainable, while others emphasize the problems that arise when government programs to address distributive inequalities compromise individual liberty or processes of economic growth.

It is worth recalling how the history of industrialization has been intertwined with the emergence of novel approaches in ethics. The fundamental problem in food ethics showed how versions of utilitarian ethics are notorious for their ability to rationalize structural inequalities that are offset by the benefits of economic growth. But the original British utilitarians were dogged critics of the inequality-perpetuating social order associated with a rigid class system. Aristocratic families in England and across Europe enjoyed enormous wealth and political privileges that allowed them to control decision making. With few exceptions, they were loath to undertake political or economic reforms that were contrary to their particular interest, even when doing so was manifestly in the interest of a vast majority. Jeremy Bentham's method of calculating benefits and costs and his maxim, "everybody to count for one and nobody to count for more than one,"[14] was a radical break with this system of class privilege. Views on human rights or Kantian principles of human dignity can be interpreted as attempts to preserve the egalitarian spirit of Bentham, while also countering utilitarian arithmetic's insensitivity to the distribution of benefits and costs. In the era of industrialization, human rights were closely tied to democracy and the right to participate in the process of governance.

The historical debate over the justice of distributive inequalities favoring the rich does not exhaust the ethical significance of the Brundtland approach to sustainability, however. Indeed, in calling attention to the question of distributive justice for future generations, the Brundtland Commission was expanding the scope of concern considerably. Since future generations are unborn, their right to influence decision making cannot be incorporated in democratic processes by extending an opportunity to participate. Democracy is of necessity a matter of decision making in the here and now. At best someone can represent the interests of future generations, but the idea of allowing them to speak for themselves is a nonstarter. The Brundtland Commission's definition of sustainability is a rare statement of philosophical principle by a political body, and it unswervingly acknowledges that the problem of distributive justice is not simply a question of distributing benefits among rich and poor. It is also a problem of distributing the costs and benefits of industrialization across time. The Brundtland approach to resource sufficiency implies that it would be wrong to solve the present generation's problems of food security by imposing environmental costs on future generations, especially if those costs limit their ability to produce food. But implementing a concern for future generations involves a philosophical puzzle formulated by Derek Parfit: if what we do today affects what kind of lives and interests future generations will have, there is a sense in which the very identity of future generations is affected by what we do. Parfit has argued that it is an error to think that we can harm future generations in the conventional sense, because they owe their very existence to our decision making.[15]

But the Brundtland report has also been criticized for not expanding the scope of distributive justice far enough. The Brundtland definition of sustainability is limited to the costs and benefits human beings experience. It is anthropocentric. In contrast, a generation of environmental philosophers has argued that the scope of our ethical thinking needs to be extended beyond the human species. As Chapters 1 and 5 discuss, some emphasize the fact that, like humans, nonhuman animals experience pleasure and pain, and some would go so far as to say that other animals experience a form of subjectivity that entitles them to moral rights. Other environmental philosophers note that all living things can be said to have interests or a good of their own. As such, there is a ground for extending questions of distributive justice to all living things. And some environmental philosophers have argued that species, ecosystems, and possibly planet Earth itself have a form of intrinsic value that deserves moral consideration. Any of these views pose a problem for the assumption that we have adequately distributed the costs and benefits of industrial production when we have only considered the rich, the poor, and future generations of human beings. They suggest that when we ask, "Sufficient for whom?," we should be considering how any human practice (including food production) affects a much larger class of morally significant beings. At the same time, few who defend such a radical reinterpretation of distributive justice would go so far as to claim that these nonhumans deserve exactly the same kind of moral consideration as a human being. This kind of environmental ethics accepts the view that ethical questions for sustainability take the form "Enough for whom?" at the same time that they challenge the Brundtland Commission's assumption that the whom would be limited to human beings.

In summation, the resource sufficiency view of sustainability comes equipped with a set of ethical questions that, while difficult and challenging, are widely recognized by philosophers, environmentalists, and policymakers. There are large literatures devoted to these problems, though relatively few authors have considered how they relate specifically to food ethics. As the previous chapter on the fundamental problem in food ethics illustrates, there may be some particular philosophical challenges that have been overlooked. Yet overall it is fair to say that food is just one good among many that are inequitably distributed in the present world. Health care, housing, energy, and basic human services are also targets for distributive justice. There are questions about how access to and use of these goods are distributed among people as they are identified by categories of wealth and poverty, on one hand, or race, class, and gender, on the other. There is also a set of questions about how production and utilization of all these goods affect both the opportunities of future generations and the prospects for a wide array of nonhuman entities, some living and some not. There is an important sense in which these are not really questions in *food* ethics, because there is nothing that distinguishes food from all the other goods that are subject to distributive inequalities. I call this the *industrial philosophy of agriculture*: agriculture is just another sector of the industrial economy. While the goods (e.g., food and fiber) produced by agriculture are subject to all the ethical debates that arise in connection with distributive justice under any of its many guises, there is nothing particularly special about agriculture compared to any other sector (e.g., health care, energy, manufacturing, transportation) in the industrial economy.

Agrarian Philosophy:
A Philosophical Interlude

In contrast to the industrial view, advocates of ecological integrity in farming or social sustainability tend to see agriculture as having exceptional features that make sustainable agriculture a problem of special significance. Because these views emphasize the singular characteristics of farming and other forms of food production, I lump them together as expressing an *agrarian philosophy of agriculture*. Farming is special; we learn something unique and important about sustainability by considering the ecological and social dimensions of agriculture very carefully. The ways in which farming is thought to be special vary, and a few of them will be discussed in the following. Notice, however, that in claiming a distinctive role and status for agricultural production, the functional integrity approach to agricultural sustainability will bring a number of unusual and in some cases long-forgotten philosophical ideas to the foreground.

Perhaps surprisingly, G. W. F. Hegel's lectures on the philosophy of history provide an entrée to the agrarian point of view. Hegel identifies four philosophically significant stages in the history of civilization, and associates them with the geographic regions in which they originate. History begins in Africa with the arrival of tribal societies and progresses to Asia, where the first diversified civilizations appear. The crucial stages are Greece (by which Hegel means the Greco-Roman world) and Germany (by which Hegel means Europe), where successively more complex forms of social organization support the emergence of social norms and the accompanying mentality or spirit associated with enlightenment and human progress. Students who attended Hegel's lectures on the philosophy of history felt they provided an accessible introduction to Hegel's notoriously abstruse philosophy of absolute spirit, but that may have been because they shared certain Eurocentric prejudices with Hegel about the nature of progress and rationality. A contemporary student must suspend the temptation to fault Hegel not only for views that we would take to be racist but also for a view of historical events that cannot help but look eccentric to a twenty-first-century mind. To the extent that one is able to do this, one can appreciate that Hegel tells a story about how agriculture could be thought exceptional that is fascinating in its own right and exemplifies some key elements in a functional integrity view.

For Hegel, there is no development in Africa because life is too easy.[16] Food is plentiful. It literally grows on trees. While Africans form tribal parties for scavenging and hunting and become involved in conflict over territory, they do not need to develop complex forms of civilization. That need arises in Hegel's Asia, which includes ancient Egypt, and can be illustrated by considering Egypt's complex system of agricultural production. Egypt is both cursed and blessed by the Nile River, which brings a flush of soil-nourishing nutrients down from the highlands in an annual flood that covers much of the country. The Egyptians converted the curse of the floods into an unmitigated blessing by building large stone and earthen works to retain the Nile's floodwaters and by learning how to release the waters at carefully planned intervals. The Nile provided reliable water and fertilizer to the entire length of the Nile River Valley, allowing the Egyptians to support a significant portion of their labor force NOT in agriculture. Most of them were slaves, and when they were not building or maintaining the irrigation works, Egypt's leaders found a way to keep them busy in the construction of the monumental architecture that we know today. Others were detailed to the army, which was used to control Upper and Lower Egypt. Still others were in the priesthood, but in addition to some ceremonial functions, the Egyptian priesthood was very busy managing the release of water for downstream farming as well as the harvest, storage, and distribution of the crops.

Hegel does not overstress the complexity and environmental sustainability of the Egyptian agricultural system, but a technically sophisticated contemporary analysis by Marcel Mazoyer and Lawrence Roudart notes that the Egyptian approach to the management of natural resources and its key social institutions endured for over seven hundred years.[17] That would be a record of endurance (if not sustainability) that no other known system of agriculture could begin to match. Of course, the Nile's annual influx of natural fertilizer was a big help, and reliance on slavery makes the Egypt of the pharaohs a nonstarter for those who associate sustainability with social justice. Nevertheless, the stability and resilience of the system is a lesson of sorts for those who would ponder how future generations will feed themselves. Hegel does note that the members of the Egyptian priesthood attained a perspective on the complex interworkings of their socio-technical system for food production that no one would have been likely to achieve in hunter-gatherer tribes. The central management perspective involved a vision of how the respective parts fit together and gave the Egyptian leadership a recognition that certain things had to happen at certain times if they were forestall the entire operation's collapse. While no one in Hegel's Africa understood the forest or savanna ecology that produced nuts or berries to scavenge or game to hunt, the despotic leader in Egypt can see what is needed in order for that country to survive. Hegel believes that *only* the despot is in a position to know this, and this is the crucial limitation to the Asian stage of human history.

It is in Hegel's treatment of Greece that the story gets really interesting. Hegel's history of Greece begins with the same geography and climate themes that we see in his accounts of Africa and Asia. The Peloponnesian Peninsula is mountainous but dotted with fertile valleys that are well watered by rainfall and snowmelt. The topography is well suited to a style of farming that incorporates annual grain crops sown in the valley bottoms, complemented with tree and vine crops that hold erodible calcareous soils on sloping lands. Sheep and goats can move through this system, wintering on crop stubble, eating the early grass in orchards, and summering on high mountain pastures, fertilizing all these lands with naturally deposited manure through each season. The system is suited to management by a household labor force, as one can move from one farm task to another by season. Trees and vines can be tended both before and after crops are planted; olives are not typically harvested until late autumn or winter. Children and teenagers can tend the livestock. All of the crucial decisions take place at the household level. Not only is there no need for the central management and hierarchical organization of Egyptian agriculture, the close relationship between decision making and the point of production in the Greek system encourages the growth of farming knowledge that is attuned to the peculiarities of a particular place.

What is more, the mountains themselves frustrate centralized decision making, especially in comparison to the ease of communication (not to mention movement of troops) made possible by transport up and down the Nile River. Greece thus evolves as a civilization composed of relatively freestanding and indepent valley kingdoms, each with an urban center where the wheelwrights, blacksmiths, and other crafts needed to support household farming are clustered. With a loose interpretation of an argument developed in Aristotle's *Politics*, Hegel argues that the household management (the *oeconomics*) of these diversified Greek farms becomes the model of governance of the *polis*, the political unit comprised of urban centers and the somewhat independent farms surrounding them. Just as people naturally develop a sense of belonging and loyalty to their family, the Greeks develop a sense of solidarity among families that becomes the basis for a norm of citizenship. For Hegel, this advances the historical development of human beings' ability to perceive themselves as part of social organism on which they depend and to which they owe allegiance. It will be followed in Hegel's philosophy of history by an increasing recognition of shared interests and common fate, as well as more sophisticated understandings of law and spirituality, as Greek civilization is succeeded by the Christian era and the emergence of constitutional monarchies.

A Hegelian picture of the relationship between Greek farming and the virtue of citizenship can be fleshed out by consulting the work of Victor Davis Hanson, a contemporary historian of ancient Greece. In *The Other Greeks: The Family Farm and the Roots of Western Civilization*, Hanson considers the mentality of these household farmers, the *hoi mesoi*, of the Greek city states.[18] Annual crops were vulnerable to destruction or pilfering by invading armies, as they would have been anywhere in the ancient world. But Greek farmers were less willing than others to simply take to the hills and wait out a siege. They could not tolerate the destruction of their tree and vine crops—crucial to their farming system—because, unlike annual grains, they could not simply be replanted once the invaders had gone. Trees and vines represent a lifetime of labor: the olive is a particularly slow-growing tree that can produce for centuries. Harvests lost to marauding armies would be foregone for years or even decades. According to Hanson, this made Greek farmers especially willing to defend their farms and to demand forms of governance that would ensure their security. The eventual result was the phalanx, a military form that depended on each soldier's absolute confidence in his comrades. The phalanx demanded a courage born of every individual's recognition that the success of their farms was dependent on their military prowess and the conviction that every person in the formation shared an appreciation of the desperate cost of failure. It was this mentality that underwrote the Greek conception of citizenship and that permitted Greek city states to field troops of farmer-citizens that were able to rout foreign armies drawn from slaves and mercenaries.[19]

As with Hegel's anthropologically naïve understanding of Africa, the historical accuracy of this account may be less important than the way it allows us to paint an alternative portrait of sustainability. Environmental sustainability depends on the continuous reproduction of fertile soils, a reliable source of water, and the resilience of cultivars in response to the vagaries of weather, pests, and diseases. But Hegel's philosophy of history suggests that a *society* is sustainable when its institutions, its laws, and its morality both reinforce and are reinforced by a pattern of practice for securing material sustenance. Both Egypt and Greece were blessed with an agroecology that was sustainable in the first sense, but Greek agriculture permitted the emergence of a social ecology as well. It linked household farms with more specialized craft workers through mutually reinforcing practices that gave a much larger portion of the population a vantage point from which the workings of the social organism (and its dependence on the natural environment) could be perceived. Both Greece and Egypt depended on slavery for a fair portion of the grunt labor. These systems were far from perfect. Yet Greeks were able to see their societies as functioning wholes, and this vision was closely tied to the Greek conception of moral virtue. One could not be virtuous in a corrupt, ill-functioning society, and the continued practice and reproduction of virtue was crucial to the institutions (including and perhaps especially the household farms) that gave people the strength to resist corrupting influences. Crucially, agriculture is far from just another sector of the Greek economy. Agriculture is the ultimate source and underpinning of the social forms that most characterize the distinctive Greek way of life.

The Ethics of Functional Integrity

The first point to take from this tangential interlude is that an agrarian picture of agriculture suggests ethical questions that *simply do not come up* when one takes the resource sufficiency perspective on agricultural sustainability. Hegel's philosophy of history suggests that for the Greeks, at least, the organization of the food system was intertwined with the practices that allowed Greek city states to reproduce the structure of their society from one generation to the next. Agriculture thus has a value that derives from its function or role in the reproduction of basic social institutions, including the characteristically Greek form of morality. Or, to put it another way, agriculture plays a role in the sustainability of Greek society and cultural life. It is certainly true that agriculture is also playing a role in feeding the Greek people. It is logically possible to ask the basic resource sufficiency questions about the Greek agricultural economy. Indeed, asking questions about enough and for whom might get us around to challenging the Greek dependence on household slaves, and it is ethically important to ask such questions. Yet these questions will *not* get us to the systems perspective that allows us to see how Greek farming practice contributed to the social organization of the *polis* and the functional significance of Greek virtues such as citizenship or patriotism, courage, and perhaps even *sophrosyne* (discussed in Chapter 3).

An agrarian philosophy contends that the food system, including the characteristic organization of farm production, is fundamental to the functional integrity of a civilization or way of life. Farming and foodways are imbricated within the patterns of practice that reinforce cultural norms or organizational institutions that hold the society together. In the Greek case presented jointly by Hegel and Hanson, these patterns are tied closely to the social ability to resist threats of military violence from outside sources. But it is possible to conceptualize a number of other ways in which food systems might fail to perform functional roles. Some of these are obvious and would be recognized by anyone evaluating a food system from the perspective of food availability: a plague of locusts, a drought, or permanent soil loss through erosion or declines in soil fertility—all of these threats lead to system failure in terms of the ability to provide adequate food. At present, a number of concerns about the robustness of the world food system are associated with climate change. But other threats and failure modes are more subtle. Undercutting dietary patterns or food knowledge that is crucial for public health might be one example, while the decline of rural communities that was the focus of the authors Douglass associated with social sustainability is another. Although a full discussion is beyond the scope of this book, some themes in functional integrity can be applied comprehensively to sustainability.

In general, a system's ability to resist perturbance from an external threat is characterized as robustness, while its ability to recover from perturbance is referred to as resilience. In the case of modern infrastructures such as the electric grid, the rail or highway transport system, or a system for supplying fresh water to urban centers, robust systems are those that resist failure as a result of natural disasters (e.g., a tornado or a hurricane) and human threats (in today's world, primarily a terrorist attack), while resilient systems are those that can be brought back on line quickly, once they have failed. But there can also be internal threats to infrastructure systems: design failures, poor maintenance, and human errors. Robustness and resilience apply to internal threats, as well.

Thinking of an infrastructure system in a broader perspective, one could say that when political decision makers fail to provide the financial resources to design, maintain, or staff an infrastructure, there has also been a system failure—a breach of functional integrity. This might happen when governments become corrupt, or when people become so ignorant of their dependence on infrastructure that they continually defeat the taxes needed to pay for them. Recall how this phrase echoes Jefferson's praise of farming cited in Chapter 4. This kind of scenario (realistic enough in contemporary democracies) speaks to the kind of sustainability that Hegel's philosophy alerts us to.

A social system in which defense, governance, and farming are intertwined in the manner that the Hegelian picture suggests can fail when the incentives that give rise to the citizen-soldier-farmer mentality are threatened. This is exactly how Victor Davis Hanson reads the history of ancient Athens. He argues that Athens became a special case because the development of sea power laid the basis for trading interests that lacked this land- and place-based conception of citizenship. The erosion of Athenian virtues tied to the farming class (the *hoi mesoi*) became, in Hanson's view, the back story to expansion of the Athenian sphere of influence, eventually resulting in the Peloponnesian War between Athens and Sparta. Hanson reads the trial of Socrates, Plato's conception of justice and law, and finally Aristotle's ethics as referring elliptically to a policy conflict between those who believed that departing from the agrarian model was the root cause in a series of events that had undermined the functional stability of Athenian society.[20] One might read Machiavelli's account of decline in the Roman Republic in a similar way: a series of bad policy decisions undermine the mutually reinforcing links between Roman institutions of governance and the Roman citizen-soldier-farmer's ability to support the army, both financially and as a participant. Tracking and debating such interpretations of classic philosophical works connects food ethics to the events that many Western philosophers associate with the very origins of their discipline. But while Hanson's account of "the roots of Western civilization" represents a tantalizing way to integrate agrarian themes into philosophy, it is also a tangent that we are best advised to resist in the present context.

The ethics of functional integrity valorizes the robustness and resilience of systems. We might also add adaptive capacity. Biological species and populations of organisms exhibit the ability to change their traits. The adaptation of peppered moths during the period of industrial expansion in England is an often-cited example. As factories belched smoke into the atmosphere, lighter colored moths became easier for predatory birds to spot when they sat on soot-stained surfaces. But the comparatively rare dark moths were camouflaged and harder for the birds to see. Naturalists observed adaptation as the population of peppered moths shifted in the predominance of light versus dark colored individuals. And, as biology textbooks note, the proportion of lighter colored moths began to increase again when environmental reforms removed the pollution that had given darker individuals the advantage in evading predators.[21] The capability for adaptive response to change in the environment must be added to robustness and resilience in order to have an adequate picture of sustainability as functional integrity.

Integrated functional systems tend to be robust, resilient, and adaptive. If we are talking about a water system or an electric grid, such traits are good because they ensure that the delivery of these utilities will be more reliable for human users. In the case of the peppered moth, many environmentalists have seen functional integrity to be in service to the population itself. But populations (perhaps unlike individual moths) do not have feelings. Similarly, if we talk about the agricultural economy of ancient Greece or the political structure of the Roman Republic, we seem to be valorizing entities that do not have feelings or the ability to care about their own fate. At this point, functional integrity starts to connect with arguments in environmental ethics that stress intrinsic value. There is some sense in which ecosystems, species, populations, and other similarly complex, self-organizing systems exhibit a kind of value or worth that transcends the usefulness that any individual—human or not—derives from them. Although there may be anthropocentric arguments for valuing the goods and services sufficiently complex systems produce, Holmes Rolston has argued that self-replicating and system-preserving aspects of these systems embody what we mean by intrinsic value itself. Our ability to perceive functional integrity is a source—perhaps *the* source—of our ability to understand value in nature.[22]

Given Rolston's view, Hegel's philosophy of history can be read as an exemplification of a long-running evolutionary pattern in the human story. As human civilizations have evolved, they have also come to embody greater robustness, resilience, and adaptive capacity. The transition from Hegel's Africa and Asia to Hegel's Greece may be something of a just-so story, but the moral point it is attempting to communicate is that human culture and individual virtues are features that contribute to the overall integrity of the social organism. We humans err when we presume that we can exhaust the moral significance of culture and virtue by identifying the instrumental usefulness of these adaptations. It is the larger adaptive trajectory and the resilience of a cultural form that provides the very source of what we, from our individual perspectives, call value. The particular preferences and ethical standards that any population of human beings happens to have at any juncture in this larger trajectory derive *their* value from the way that they reflect and contribute to the functional integrity of the larger whole. Not all epochs in human history can be regarded as progressive by this standard. Sometimes we regress. Of course, people who live during any period must rely on whatever standards and ideas are made available by the ethical discourse that is characteristic of their time, but the larger task of ethics is to envision some means of testing one's own codes against this larger and more encompassing trajectory. Early nineteenth-century thinkers such as Hegel were confident that their societies were heading in the right direction. It is difficult to be so confident after two world wars and a half century of smoldering (but deadly) religious conflict. A series of technological failures and the ominous prospects of environmental depletion and climate change also blunt our faith in progress. At the same time, from the current vantage point in history, we can see in Hegel and his contemporaries a remarkable insensitivity to the perspectives of women and non-European racial and cultural groups. If we can grasp a Hegelian understanding of robust, resilient, and adaptive growth in human history, we can also move beyond Hegel in seeing the limitations of nineteenth century idealism. Perhaps progress is not completely dead.

The last important point to draw here is that questions about the functional integrity of a system depend on how one understands the boundaries of that system. In the 1970s, the American corn crop was devastated by a disease known as Southern corn leaf blight. The cause was eventually traced to the fact that virtually all of the corn being grown in the United States was derived from the same genetic source called T-cytoplasm after a Texas variety of corn in which it had been discovered. This particular variety was uniquely susceptible to the disease, and once corn breeders began to draw on a more diverse gene pool, the threat to the nation's grain supply was averted. But one might ask, "What was the relevant system here?" Agricultural scientists tended to think that it was genotype-environment relationship between T-cytoplasm and *Helminthosporium maydi*, the fungus that caused leaf blight.[23] However, it is quite unlikely that any such thing could happen to Mexican corn farmers, who carefully manage genetically diverse open pollinated varieties of maize known as "land races," even if *Helminthosporium maydi* were present in their environment. Mexican corn farmers see the diversity of their land races as intrinsically valuable. They think of themselves as "people of the corn" with a sacred responsibility to assist the plants under their care in reproducing that diversity. Their corn is not just useful for food; it is the lynchpin in the socioecological system that defines their very sense of identity. Perhaps we don't grasp the appropriate system boundary for evaluating the sustainability of American corn production until we include the incentive structures and cultural values that led hybrid seed companies to utilize T-cytoplasm, or that led American farmers to buy those seeds in copious quantities.[24]

The point that I am trying to hammer home here is that value-based questions of enough and for whom do not lead us to question the values that may lie hidden in the way we conceptualize the relevant system of reference. These *are* values, but they are values that reflect the way people develop and work within a given understanding of their world and their place in it. They are values that reflect key social roles: consumer, businessman, scientist, activist, or public servant. They define the worlds in which people working from any of these perspectives make decisions and operate. An agrarian view suggests that the role of farmer is uniquely positioned to grasp key system interactions and to appreciate key vulnerabilities. [25] A contemporary agrarian might well argue that the decline of this role in modern society is a key vulnerability in the functional integrity of modern industrial society. Wendell Berry is the contemporary agrarian who is most prominently associated with this kind of argument, though Victor Davis Hanson might arguably be another.

Some Contemporary Debates

Chapters 1 and 2 discuss the emergence of a contemporary food movement. As noted, a naïve or common-sense view of food ethics begins with the presumption that individuals can make food choices that will help promote beneficial outcomes. This model fits well with ethical theories that portray *all* of ethics as a matter of making the choices that bring about the best outcomes, once all factors have been considered. It is also quite adaptable to rights-based approaches in ethics, which emphasize the way that our choices—what economists call our opportunity set—must be constrained so that we do not even consider those options that violate the rights of others when we begin to weigh the costs and benefits of each possibility. These models can be expanded to consider not only individual dietary choices but also society-wide policy choices. It becomes possible to undertake an ethical inquiry into food that focuses not only on the way our own diets are tied to better or worse outcomes but also on the outcome produced by policies governing food production and distribution, including standards for animal care or environmental regulations.

The food movement has given rise to a series of "new foods," each of which has been thought to serve as a response to ethically unacceptable consequences associated with the industrial food system. As Ann Vileisis argues in her book *Kitchen Literacy*, this new consciousness of ethics was at least partially a response to the way consumers willfully allowed themselves to patronized by experts and food industry firms well into the 1950s. Gradually, snafus over the regulation of food additives or the residue of agricultural pesticides started making their way into the news. The public's first reaction was simply to prefer "natural" foods to all these things that were being added by farmers and food processors. But the enthusiasm for naturalness played out as the food industry learned to exploit the inherent vagueness of that word. Natural foods were eventually displaced by organic foods, which have in turn given way to local foods in the wake of popular media reports on large organic farms' and food industry firms' adoption of organic methods.[26]

The list of hyphenated foods has now expanded to include cage-free, GMO-free, gluten-free, bird friendly, dolphin friendly, slow, free-range, artisanal, minimally processed, grass-fed, and (of course) sustainable, in a seemingly endless succession of qualifiers to the ethical diet. But *does* purchasing natural, organic, humane, fair-trade, and local foods actually achieve the desiderata of an ethical diet? The debates can be pretty contentious, and they are excruciatingly detailed. As such, it may be useful to consider a somewhat stylized case. A fairly large number of philosophers who have decided to enter the fray on food ethics have advanced arguments that go something like this:

> Feeding grain to animals is inefficient because it takes between two and six ounces of plant protein to produce an ounce of animal protein in the form of meat, milk, or eggs. Humans could have less impact on the environment by eating the grains directly rather than feeding them to animals. What is more, animals emit greenhouse gases that are contributing to global warming. Thus, given the animal suffering associated with feedlots and the factory farms that have been developed for

industrial production of pork, eggs, and poultry meat, food ethics requires that we at least consume animal products from small, local farms, if not shifting to a vegan or pescatarian diet.

I have run a number of things together to avoid dragging readers through an overly detailed discussion of all the variables and variations on this theme. Nevertheless, many contributions to contemporary food ethics adapt the argument Frances Moore Lappé proposed initially in the first edition of *Diet for a Small Planet* back in 1971.The arguments conform roughly to the above summary.[27]

The climate ethics component of these articles often cites a report from the FAO entitled *Livestock's Long Shadow*. Although advocates of an environmental or local "food ethic" usually cite the report in support of the view that climate impacts are prominent among industrial livestock production's environmental effects, this reading runs deeply counter to the intent of the report's authors and is not supported by the facts and figures summarized in the report's analysis. What the report actually shows is that pasture-based systems for producing meat, milk, and eggs are inefficient users of land and more harmful contributors to greenhouse gases than are intensive CAFOs, where animals are fed on grains from industrial monocultures. The FAO report makes projections of livestock's continuing impact on the environment as more and more consumers in the developing world increase their consumption of animal protein, and concludes that a failure to introduce Western-style industrial production systems in the developing world (where intensive systems were still predominant when much of the analysis was done) will have devastating environmental impact.[28] In fact, there have been numerous scientific studies showing that however much meat, milk, and eggs human beings decide to eat, the confinement systems that utilize industrially produced animal grains win in a comparison with extensive animal production systems, at least when the alternatives are evaluated on a unit-by-unit basis. That is, if one considers the per pound or per ounce environmental impact of animal production, the scientific literature indicates that the industrial systems are able to achieve efficiencies in total land use, total energy consumption, and total greenhouse gas production when they are compared to traditional pasture-based production.[29]

Indeed, the entire movement toward local food systems may have been built on an idea that was initially floated more as a thought experiment than as a factual claim about the environmental impacts of the industrial food system. In 2001, a team of Iowa State University researchers issued a report entitled *Food, Fuel, and Freeways* in which they presented the idea of a "food mile" as a heuristic for considering the contribution that transport of food makes to carbon emissions. The idea of a food mile had been around at least since a 1969 study by the US Department of Energy used it as a proxy for estimating the amount of energy used in transporting food. Yet the Iowa State report (which did not cite peer-reviewed claims and was not represented as more than a speculation) quickly captured the imagination of food-system activists and environmentalists—just the sort of people who were advocating for local food ethics. The idea that "eating locally" by shopping at farmers' markets or limiting one's diet to items produced within a one-hundred-mile radius could make a significant contribution to the reduction of greenhouse gas emissions became a key rationale for the transition from a food ethic of "eat organic" to one of "eat local."[30] But once subjected to a life-cycle analysis, the food mile balloon began to pop. When the fuel for multiple trips in a car or truck for frequent shopping is compared to that of a long-distance semitrailer or—even more dramatically—ocean-going freighter carrying tons of grains, fruit, or vegetables, a food miles argument simply does not hold up.[31]

Readers who question this result should read the sources cited in the footnote carefully. At least some of them are written by authors with a strong philosophical commitment to sustainable agriculture. This does not deflate the case for local food entirely. Perhaps it would be better to say that the question just becomes much, much more complex. My point here is not to debunk the overall thrust of a food ethics that proceeds under the assumption that people can further environmental ends through their food choices. To some extent, it is impossible to deny that promoting more environmentally sustainable consumption choices, along with more healthful or socially just choices, is what food ethics should try to achieve. It is important to notice that none of the studies I have cited *do* deny that comprehensive norm; what they dispute is whether some specific and widely advocated rules of thumb for dietary choice actually promote more environmentally sustainable consumption. Sorting the arguments out requires a review of the assumptions that are made in efficiency calculations, projections of food demand, measurement of carbon and other pollutant emissions, and the phases of a life-cycle assessment. And if we work through issues such as the true health benefits of organic food, the actual risks of pesticide use, the effect of producing biofuels on the global food supply, or the enormously contentious debates over genetic engineering, we find ourselves wading through one episode of examining assumptions after another. This kind of sorting out is a blend of epistemology, philosophy of science, and straightforward (if complex) factual analysis. While it is important to dig into these questions (and the concluding chapters do a bit of that), they are not questions that usually turn on ethical considerations. Food ethics morphs into something we might call "food epistemology."

The Moral of the Environmental Food Ethics Story?

Leaving the contemporary debates at this juncture misses a key environmental ethics debate that we *should* have, however the factual questions are resolved. As I have argued at greater length in *The Agrarian Vision*, the naïve food ethic that focuses intently on food choices and their economic impacts operates almost entirely within the assumption that agriculture is just another sector of the industrial economy. Like mining, manufacturing, transportation, or even health care, there are choices to be made that affect the overall consumption of resources and the risks we bear as a result of polluting activities (such as carbon emissions). There is

really nothing ethically unique about food consumption in considering these choices, save for the fact that we make food consumption choices far more frequently than we buy washing machines,
build houses, or fly in airplanes (but not necessarily more frequently than we ride in automobiles). This approach essentially *is*
the resource sufficiency approach to sustainability. By evaluating
our food choices under this kind of ethical paradigm, we are *ignor-
ing* the systems perspective that is brought into consideration by
understanding sustainability in terms of functional integrity.

A functional integrity approach would ask us to consider food
ethics first in terms of our food system's organization and second in
terms of how that organization suggests that each of us has a particular place in the system. The perspective of consumer choice implied
by the industrial food system suggests that we might consider how
what we choose to eat affects people, animals, and the environment
through the complex economic causality of markets and trade, but
it does not really bring us to consider whether a system that encourages us to think of ourselves as consumers is itself ethically desirable.
How else might we think of ourselves? Hegel's citizen-soldier-farmer
is one alternative. The practice of farming in a *polis* gives rise to conceptions of virtue, solidarity, and interdependence that illuminate
the mentality of the *hoi mesoi*. But while the citizen-soldier-farmer is
a coherent alternative to the consumer mentality, I would not want
to be read as saying that we should emulate the *hoi mesoi*. Greek society had singularly unattractive features as well. Nonetheless, Hegel's
citizen-farmer *is* a way to illustrate how food systems were once configured in ways that penetrated deeply into the structure of human
practices, institutions, and moral ontology. In fact, an adequate environmental ethic for food will require that we place functional integrity and resource sufficiency in dialogue with one another. Questions
of enough and for whom must be asked, but so must questions of
how our systems, our environments, are reproduced over time—and
whether they should be.

If you take an environmentalist/utilitarian view of food
ethics—if you think that food choices should be evaluated by their
impact or consequences—then you are very likely to be taking

some version of a resource sufficiency view on sustainability. If you follow the thinking of the Brundtland report, you will include impacts on future generations, as well as on poor people living in less-developed parts of the world. You might even be a radical environmentalist/utilitarian who wants to include the well-being of nonhumans, and if you are a *really* radical consequentialist, you might even think about the well-being of plant species or eco-systems. You are going to have to negotiate some tough tradeoffs among all these different impacts, but in every case, truly more efficient systems of food production are going to come out very well. Like Laura Westra (discussed at the beginning of this chapter) you might want to build barriers between nature and human society, and agriculture might become a buffer zone that helps to protect wild ecosystems from the harmful impact of industrial practice. But the more efficient that buffer zone is in producing consumable foodstuffs, the more room we leave for nature. The hard questions then become questions of *measuring* efficiency and impact. The ethics part is easy, except when we face an unavoidable trade-off.

This view neglects to recognize the way *who we are* as individuals and as cultural groups is itself the product of an organism-environment interaction. Human beings are complex organisms capable of reflecting the experience of long-dead generations into their current experience as well as projections for generations yet to come. There is a profound sense in which our history is part of our environment. The functional integrity view requires us to grasp this larger sense of what it means to form values or an identity within an environment, and then to think of sustainability in terms of whether the key functions that produce human capabilities (as well as those of other species) are being reproduced in our children. What is more, our ability to think historically and environmentally means that we can even ask whether we ourselves have retained the capabilities that preserved human civilizations in the past. We can think about the robustness and resilience of our current institutions, and if we find them wanting, we can think about how they might be adapted in ways that recover corrupted modes of functioning.

Perhaps the best ethical argument for local food systems—buying from farmers at farmers markets, eating in season, and joining a community-supported agriculture or local food co-op—is *not* so much that the food choices we make have ethically better outcomes. Perhaps the reason to advocate for sustainable agriculture is to encourage people on a journey that helps them realize their place in the world, that brings them to an awareness of the systemic reverberations of their eating practices and to question whether there might be ways to more frequently remind themselves of the vulnerability, contingency, and uncertainty of the systems—social *and* ecological—on which our ways of life have been built. I do not go beyond saying "perhaps" here. To think about sustainability in terms of functional integrity is not to ask whether we are leaving our children "enough." It is to ask whether the kind of environment that made us who or what we are today can and should endure. It is to ask whether the sociocultural environment that produced *us* is the one that we really want our children and grandchildren to have. Such an inquiry questions whether our social world—and especially our food world—is truly the kind of environment that we would want to shape the character and habits of our children.

That may be the hardest kind of question an environmental philosopher can ask. What is more, it is a line of questioning that smudges the distinction between trying to understand how the production and consumption of food affects the well-being of others (including non-human others) and the *ethos* bound up in foodways—the way our eating habits spill out into habits and practices that make for distinctive cultures. Perhaps contrary to the remonstrations of Chapter 1, we are what we eat, after all. Yet if the journey of self-reflection invites us to probe the sustainability of our food system more deeply, I would not abandon caution against imposing this burden of reflective deliberation on everyone. The questions of resource sufficiency may be philosophically shallow, but "Enough for whom?" remains a vital question. And the demand for distributive justice should temper any enthusiasm we might have for integrative food systems that engender virtue and insight, but only for the few.

Chapter 7

Green Revolution Food Production
and Its Discontents

Do wealthy people and their governments have an ethical obligation to moderate their qualms or concerns about new agricultural or food technology in virtue of its potential for alleviating hunger and malnutrition? The question may appear arcane at first, but it follows naturally from the caution with which the previous chapter ended. What is more, it has become relevant to a surprising number of policy issues that are central for food ethics. So-called GM (or genetically modified) foods have been developed through recombinant DNA techniques for introducing new traits into plants, microbes, and animals. Several applications of this technology have been incorporated into food and fiber crops that are grown by farmers in the United States, Brazil, Argentina, and a number of other food-exporting nations. In practice, only two general types of GM crops have been widely adopted so far, and both have increased food security, improved farmers' livelihoods, and reduced harmful environmental impacts. Bt crops incorporate genes from the *Bacillus thuringeinsus* bacteria. The genes produce proteins toxic to caterpillars but not known to have effects on vertebrate species. These toxins have traditionally been derived by culturing the *Bacillus thuringeinsus* bacteria, and this method of using Bt is a mainstay of organic farming. In GM crops, the toxin is produced within the plant itself. The other main type of GM technology is herbicide-tolerant crops. Both types have been developed for a wide array of crops, though they are still used primarily in just three: corn, soybeans, and cotton.

GM technology (sometimes rendered as "GE" for genetically engineered, and sometimes as "GMO" for genetically modified organism) sparked a firestorm of controversy in the late 1990s. Biotechnology's potential to address food shortages and nutritional deficiencies became a frequent talking point among its advocates, especially in debates over proposed shipments of GM maize for use as food aid and over the prospects for addressing Vitamin A deficiencies with a genetically modified variety popularly known as golden rice.[1] The most poignant episode arose when several African nations refused emergency shipments of food from the United States in 2002. There were a number of rationales offered for the refusal. Some African leaders questioned the safety of the food, despite assurances that Americans had been eating GM crops for five years. However, the potential impact on African exports may have been more influential. Farm leaders expressed the fear that Europeans would seek new suppliers if they had even a remote suspicion that African production was contaminated by GM varieties.[2] Though fascinating in their own right, the details of this episode would take inquiry into the question posed above far afield. The point here is to examine the underlying ethics behind various reactions to novel food technologies, especially in light of current and future global demand for food.

The Moral Issue Unpacked

The global food system has been thoroughly transformed since the industrial revolution, with the introduction of new technologies for production, transport, processing, and distribution of foods. The nineteenth century saw the introduction of refrigerated boxcars that made it possible for livestock to be slaughtered in Chicago and then shipped as sides of beef or pork to New York for delivery to local butchers. As unexceptional as this may seem to contemporary eyes, there was no intrinsic reason for a member of the consuming public to believe that refrigeration was an ethically acceptable way to handle food. In fact, the new technology created a system in which food passed through many unseen hands before going into the mouth of a consumer. This created new opportunities for the adulteration of foods. As noted in Chapter 1, Upton Sinclair's novel *The Jungle* was intended as a comment on food justice, but it was received as an exposé of the food safety risks associated with the meat industry. It was an era more committed to the inherent goodness of technological progress than the present, but the first decades of the twentieth century saw numerous attempts to tarnish the reputation of industrial food technologies on grounds of safety and aesthetics. The most well-to-do consumers may have been among the last to take advantage of canned vegetables or packaged meats.

Food ethics suggests that these well-to-do consumers were fully justified in deciding not to eat these new products, but there are other ethical points to consider. One can imagine a scenario circa 1900 in which better-off consumers lobby both government and the retail end of the food supply chain to reject the late nineteenth century's hot new techniques for the preservation of food through refrigeration. They might have been successful in keeping these products off the market, but who would have been the worse for that? Arguably, the working poor would have suffered most. Because they spend the largest portion of their income on food, they stand to benefit the most when new technology either lowers the price or stabilizes the supply of food. People who can afford to be picky eaters have a moral right to reject these new techniques for food preservation, but it might be unethical for them to create barriers that would prevent the poor from enjoying the benefits of the technology. The general ethical problem requires us to think beyond the question of what we ourselves want to eat.

The ethical issues are inherently tangled up with some matters of fact. *Are* the new technologies safe, and what level of confidence in their safety is needed to permit them? If one is confident that the new technology will not cause an appreciable increase in food-related injury, it appears that the well-off are high-handedly allowing their own preferences to dominate the poor's interest in cheaper food.[3] But what kind of preferences are we talking about? Perhaps they are merely aesthetic preferences: the well-off prefer to get fresh meat and vegetables from local suppliers because they like the taste. Perhaps the well-off feel a sense of solidarity with their longtime trade partners. Or there may be risk preferences at work: even if the evidence of safety is strong, the well-off may still prefer not to take even the smallest chance with food safety. Each of these possibilities suggests a somewhat different way to approach the ethical issues. If the issue is simply taste, food ethics suggests that wealthy people are pursuing a relatively unimportant interest at the expense of more fundamental needs. On the other hand, one might argue that the poor should not be placed in a position where they must put *their* health at risk by consuming unsafe food. The idea that one's willingness to eat risky food should be subject to market forces will strike many people as ethically problematic.

Of course, poor people are routinely placed in exactly this position. Jane Addams, the social reformer active at about the period of our imaginative scenario, was moved to commit herself to a life on behalf of the poor when she saw indigent Londoners fighting for the rotting scraps of food deemed unsalable on the cash market.[4] A utilitarian ethic would enjoin us to emphasize the most likely outcome from restricting new technology. It might be nice to suppose that a "right to food" would preclude poor people from exposing themselves to food-borne risks, but is this the most likely outcome from banning refrigeration? No. It is more likely that the status quo observed by Jane Addams will continue. If refrigeration technology is allowed (the policy that was, in fact, implemented), the net result is a larger supply of unspoiled meat and produce. This benefits the poor twice. Better food is available to them, and the larger supply of unspoiled food lowers the price. And this holds even if the process of implementing refrigerated transport requires one to accept uncertainties such as the potential for adulteration or unknown hazards. It *is* important that all the hazards and probabilities have been accounted for in a utilitarian perspective; it is not as if we should take the safety of a new technology on faith. But given reasonable confidence in the safety of new technology, it will turn out that well-off people should probably stifle their doubts. The poor are taking ethically unacceptable risks whether the new technology is used or not. The question is whether the risk of the new technology outweighs the risks of injury or starvation that poverty creates on a routine basis.[5]

Seen in light of this historical thought experiment, the next generation's food technologies (including GM crops, synthetic meats, and nanotechnologies) are similar. Concerns about new technology are countered with an argument stressing the technology's potential for alleviating a morally compelling need. Within agriculture, similar arguments were voiced in debates over Green Revolution projects that introduced chemical fertilizers and pesticides in the developing world. I accept the premise that no one should be forced to eat something that makes them queasy or uncomfortable. The discussion that follows focuses on the overriding question of when and whether such feelings about new food technology could rise to the level of an argument against allowing them to be used at all. This collapses a number of issues that could be pulled apart and considered separately. How should we assess the safety of novel food technologies, and of GM crops in particular? How should we evaluate when and whether a new technology *really does* have the potential to help the poor? What should we think about the Green Revolution, anyway? And finally, how do we evaluate all the conflicting claims that are made about a new food technology like agricultural biotechnology? Pulling these issues apart would yield a logically crisp and analytically more precise treatment, and some of the safety questions *are* discussed in Chapter 8. But a detailed discussion would not be a form of food ethics suitable for everybody because it would also get boring.

Nobel laureate Norman Borlaug (1914–2009) was a particularly vocal and prominent defender of both Green Revolution and GM crops. Borlaug never hesitated to suggest that those who voiced concerns acted immorally because the needs of the hungry override the concerns of the wealthy.[6] Likewise, advocates of stem cell research cite its potential to be used in therapies for devastating diseases as a counter to those who have qualms. Indeed, this general pattern of argument may be so common as to constitute a problem of general interest in the ethics of technology: does the potential for morally compelling benefits to the needy override less compelling concerns expressed by people who do not share this need? In keeping with a focus on hunger, I briefly review some of the main arguments that are used to characterize the moral signifi-cance of hunger and malnutrition as global problems. I also make a selective review of some arguments behind qualms that have been expressed about GM food. I then connect the two lines of argu-ment in which the claims of hunger are pitted against those of the queasy, concluding with a brief discussion of the ethical implica-tions for ongoing debates.

The Ethics of Hunger and Green Revolution Development

Although global hunger has, in some sense, been a topic of phil-osophical reflection at least since Thomas Malthus's *Essay on a Principle of Population* in 1798, it is really a topic of the last forty years. Philosophers began to write on the basis for global obliga-tions to ensure that the world's hungry are fed in response to a rising general awareness of hunger that had multiple sources and themes. One theme was certainly the Green Revolution itself, which in the early years was presented to the public as a techno-logical fix for hunger.[7] Another was the political debate over foreign aid, especially in connection with United States Public Law 480, which established the program for concessionary sales of US grain to countries experiencing food deficits. Private charitable appeals served as a third source of public awareness about the hunger of distant peoples. Children in the 1950s and 1960s were encouraged to "trick or treat for UNICEF," and organizations dedicated specifi-cally to hunger relief began to solicit funds. All of these sources suggested implicitly that addressing distant hunger was a morally good thing to do, though they did not examine the ethical basis for this suggestion.

As Chapters 1 and 4 discuss, Peter Singer's 1972 article "Famine, Affluence and Morality" gives us a philosophical analysis of world hunger. Singer argues that even moderately well-off people have a moral duty to limit their consumption of "luxuries and frills," so that they may devote their discretionary income to "those in dire need"—famine relief, more specifically. The key moral premise in Singer's argument is that "if it is in our power to prevent something very bad from happening, without thereby sacrificing anything else morally significant, we ought, morally, to do so."[8] While the theoretical lines of argument differ substantially, many philosophical authors responding to Singer see the moral imperatives of hunger as a duty arising from the need to remedy inequities and injustices associated with globalization. Offering our monetary, political, and moral support to people who are developing food technologies that increase or stabilize crop yields is seldom mentioned as a way to meet one's obligations to ameliorate the extreme needs of hungry people. Yet one way to revisit the themes taken up in Chapter 4 is to consider whether we have a moral obligation to provide support for agricultural science and development policy that contributes to this end.

This reasoning leads us to a key moral hypothesis. If Green Revolution-style efforts have the potential to reduce hunger and deprivation over the long run, then people have a moral obligation to support the use of these techniques, at least insofar as they are deployed in pursuit of that end. This might mean that everyone has an ethical obligation to support to such efforts financially. Singer argues for this kind of support in his book *The Life You Can Save*, though he does not single out agricultural development assistance, much less GM crops.[9] The hypothesis might also mean that one should lend political support to government programs for agricultural development. Here I will consider the least onerous interpretation of this obligation: that people should lend their moral support to Green Revolution-style efforts. We should at least be rooting for them to succeed.

This is a potentially controversial claim today. I will not be providing a detailed discussion of the criticisms that have been mounted against the early years of the Green Revolution. The focus here is on whether the potential for helping the poor *overrides* less compelling ethical concerns that may exist concerning GM crops and other emerging agricultural or food crops. I will call this the Borlaug hypothesis. The Borlaug hypothesis holds that even if you don't see any value in applications of cutting-edge technology for food production and processing for yourself, you should still lend moral support to any technology that has the potential to help the poor. There is an empirical dimension of the Borlaug hypothesis that can certainly be questioned. It is significant that a large part of the public debate over golden rice has concerned its effectiveness as a strategy for addressing Vitamin A deficiencies in the diet of poor people, and the legacy of the Green Revolution itself will continue to be debated by specialists. However, the ethical or philosophical elements of this hypothesis may seem securely established by the argument that Singer gave in 1972. Although Borlaug is best known for advancing this view, it has been advanced by other scientists and philosophers as well. Robert Paarlberg has made a passionate defense of this claim in his book *Starved for Science: How Biotechnology is Being Kept Out of Africa*.[10]

During Borlaug's years at the forefront of Green Revolution projects, traditional breeding was being used to develop more productive crops. Borlaug deployed a strategy that breeders knew well. Take a crop variety that thrives in the climatic conditions you are interested in, and then select for shortness in order to get a dwarf version of the rice, corn, or wheat. Dwarf varieties have shorter stalks and stems, putting proportionally more plant biomass into the edible parts. Dwarf varieties channel the extra energy from fertilizer into the seeds, while ordinary varieties will grow too tall and fall over. Borlaug pioneered a method of growing one generation of plants in the Northern hemisphere and then flying the seeds to a Southern hemisphere location with similar soils and climate. This allowed him to cut the time that most breeders were taking to develop new varieties in half.[11] For the last twenty years or more, the key points of debate have concerned agricultural biotechnologies that allow breeders both to introduce an array of traits into crops that go well beyond dwarf stature and to cut the time for crop development even further.

Arguments against the use of recombinant DNA to introduce genetic novelty into food crops are also quite diverse and complex. Any summary analysis of these concerns is likely to be controversial in its own right, but most arguments can be classified into one of five main groups. The first group of arguments includes disagreements over the appropriate way to assess and manage technologically induced risks. Second is a concern that agricultural biotechnology is incompatible with social justice. Third, there are arguments to the effect that biotechnology is unnatural. Fourth, there are arguments that stress the importance of personal autonomy with respect to food choice. Finally, there are aretaic objections that focus on the moral character of the people and groups supporting biotechnology.[12] Any satisfactory treatment of these questions would require a fairly lengthy discussion to both fairly summarize the objection and work the reasoning offered in reply. Here, I review each category of criticism very briefly.

The Case against Agricultural Biotechnology: Precaution

The main argument in debate over risk has been the precautionary principle or precautionary approach. Stated succinctly, the precautionary principle holds that uncertain risks should be given great, perhaps dominant, weight in environmental and food-safety decision making. Uncertain risks are defined in contrast to known risks, which are in turn understood as risks for which both hazard and exposure can be estimated with a high degree of confidence. To clarify the difference, an anecdote may convey what many paragraphs of analytical discussion would not. At a 1999 symposium on GMOs held in Washington, DC, an officer of France's food safety agency was challenged to explain why they were applying the precautionary principle to GMOs but not to unpasteurized cheese. The challenge was intended to show that the French were being inconsistent, but the answer was simple: "We *know* that's dangerous." Because risks from unpasteurized cheese are known, the precautionary principle does not apply. In contrast, risks may be classified as uncertain when exposure mechanisms for inducing hazards are not understood, perhaps because empirical data on the frequency of hazards is lacking or because analysts may have overlooked a novel hazard that they had no basis to expect, the so-called unknown unknown. Critics of the precautionary principle have pointed out that some elements of precaution have long been incorporated into conventional approaches to risk analysis and are, in fact, reflected in regulatory decisions that have approved the release of GM crops. Other elements, such as the concern for unknown unknowns, are ubiquitous and provide no basis for viewing risks associated with GM crops as less certain than risks from conventional foods.[13]

The ethical issues can be appreciated by placing debates over precaution into historical context. Although elements of precaution have always played a role in regulating product risks, the phrase *precautionary principle* has come into use relatively recently. It was introduced in the 1990s to summarize an argument for shifting the burden of proof both for regulating chemicals and for awarding damages in product liability lawsuits. Tobacco and asbestos provide two clear examples. In both cases, the scientific evidence for harm became more and more convincing while regulatory agencies and courts of law seemed paralyzed and unable to do anything about it. Carl Cranor has shown that the burden of proof was skewed in favor of companies: People seeking regulation or damages were being required to prove that they had been harmed, despite mounting (but still not certain) statistical evidence that numerous toxic substances were tied to heart disease, lung cancer, emphysema, and mesothelioma.[14] To call for a more precautionary approach meant reversing this burden of proof, and avoiding the error of neglecting or discounting evidence that a product was dangerous.

The practice of precaution in the area of foods is older, though the phrase *precautionary principle* was never associated with food safety policies. Although refrigeration technology is now thought to be uncontroversial, adulteration was a serious problem when the FDA was created in 1906. Ann Vileisis documents how Harvey

W. Wiley fought against industry practices that are now unilaterally characterized as "unsafe," first through his post at the USDA Bureau of Chemistry and later as the first commissioner at FDA. Arguably, Wiley was taking a precautionary approach because the evidence for harm that he drew upon was certainly equivocal at the time. Vileisis also chronicles the passage of a 1958 regulatory standard that Congress mandated for FDA use in evaluating food additives. The Delaney Clause was among the most precautionary decision rules ever utilized by a regulatory agency for the evaluation of any industrial product. It required the FDA to ban every additive shown to have any cancer-causing tendency, regardless of dose and level of effect.[15]

When food safety debates moved into the realm of artificial sweeteners, the ethics of precaution became more complex. From saccharin to cyclamates to aspartame, studies showed some evidence of risk to health, yet equally strong opinion stressed the need for artificial sweeteners as a counterweight to the obesity risks discussed in Chapter 3. Largely as a result of the order in which these alternatives were considered, and certainly *not* due to any material difference in risk, each received rather different treatment in the mass media and in regulatory review. The political response was more closely correlated with media coverage than with evidence for risk. First, cyclamates were banned as called for by the Delaney Clause. Then, in 1977, Congress prevented the FDA from banning saccharin even though the evidence was very similar to that on which cyclamates had already been banned. When aspartame came along a few years later, it entered the scene as a much needed alternative to saccharin and was given less regulatory scrutiny than had been applied to cyclamates and saccharin. The ironic result: cyclamates—probably the safest of the three—were the only alternative sweeteners banned in the United States on precautionary grounds.[16]

The practice of applying precaution in food safety saw another episode in the mid-1980s when the EPA decided to ban Alar, a pesticide used in apple production. Many scientists argued that although one could not exclude the possibility of a health risk from

Alar, it was far from being the agricultural pesticide that one would choose to ban, based on the evidence. Again, the decision became newsworthy, this time as a result of savvy activism by consumer groups. Actress Meryl Streep was recruited to a campaign that stressed the vulnerability of infants drinking apple juice.[17] As in the dietary sweeteners case, a precautionary approach led to a regulatory decision that most scientists felt was irrational and unsupported by the evidence. Bruce Ames and Lois Gold summarized their view as follows:

> Regulation of low-dose exposures to chemicals based on animal cancer tests may not result in significant reduction of human cancer, because we are exposed to millions of different chemicals—almost all natural—and it is not feasible to test all of them. Most exposures, with the exception of some occupational, medical, or natural pesticide exposures, are at low doses. The selection of chemicals to test, a critical issue, should reflect human exposures that are at high doses relative to their toxic doses and the numbers of people exposed.[18]

Ethics and epistemology—the theory of knowledge and scientific inquiry—intersect in applying the precautionary principle. Biotechnology entered the scene at a point when many scientists believed that precautionary reasoning was leading to unsupportable regulatory judgments.

Environmental risk is the other area in which the precautionary approach might be applied to agricultural biotechnologies. Again, the issues get complicated and a summary treatment must suffice. The crucial philosophical question is: what defines hazard or harm in cases of environmental risk? Some defining criteria are noncontroversial. If the release of a substance into the environment makes people ill, that counts as harm. Human mortality and morbidity are noncontroversial hazards, but that is not where the matter ends. Environmental protection implies that disruption of ecological stability or damage to flora and fauna living in an ecosystem can also qualify as harm (or at least as an unwanted and adverse impact). But agriculture is *always* the intentional alteration of a natural ecosystem, and something as simple as plowing the field causes injury and loss of life to the mice, voles, snakes, and other critters living there. Any and every application of an agricultural insecticide, fungicide, or rodenticide is intended to kill organisms living in the natural environment. Although it is logically possible to object to anything and everything that kills plants and animals on ethical grounds, it is important to recognize that this amounts to an argument against food production itself. There must be some place to draw a line between an acceptable environmental impact and an unacceptable one.

The way this line is currently drawn in US pesticide policy is to make an estimate of whether the benefits in the form of food production outweigh risks to those organisms—insects, birds, mice, and voles—that you did not *intend* to kill by using the pesticide in the first place. In addition, risk assessment also considers whether systemic impact on ecosystem functions occurs. The latter might happen if you started farming in a manner that disrupted the only habitat for an ecologically important species, for example.[19] Although regulators do not always balance risk against benefits to farmers, the chance of unintended impact on nontarget species or ecosystem functions is taken to define what is meant by environmental risk in most regulatory contexts, including biotechnology. The fact that GM organisms have the potential to proliferate when released into nature would be taken into account in evaluating how likely these hazards are to materialize. A precautionary approach would mean that one should not insist upon clear evidence of harm before taking action to avoid impact on nontarget species and ecosystem functions. This does not imply that one takes precautionary action based simply on *speculation*, however, so agreeing on when one has been "precautionary enough" will very likely be a continuing debate.

There are less principled ways to define an environmental hazard, and some show up in the debate over agricultural biotechnology. The Cartagena Protocol to the Convention on Biodiversity (CBD) defines a living modified organism (LMO) as any living organism that possesses a novel combination of genetic material obtained through the use of modern biotechnology. The CBD then defines an environmental risk as any threat to biodiversity resulting from an LMO. As mentioned in Chapter 6, risk to biodiversity is somewhat ill-defined and not easily quantified. If one examines CBD guidelines for environmental risk assessment, they stress non-target organisms and ecosystem functions, just like the approach that has just been sketched above. However, the language linking LMOs and "threats to biodiversity" has been an invitation to very broad interpretations of what would count as a risk to biodiversity. For one thing, agriculture itself is a significant threat to biodiversity. It is very reasonable—indeed, I would say crucial—to think carefully about when and how agricultural production should be expanded, especially in those parts of the world that are known to be important centers of biodiversity. Yet it is a mistake to think that agriculture involving an LMO is inherently more dangerous to biodiversity than standard forms of crop and livestock production. Indeed, some people seem to think that any LMO found outside a farmer's field is not merely a *risk* (e.g., damage to biodiversity *might* occur) but evidence that the environmental damage has *already* occurred. If we speak this way, we have thrown biology out the window and are essentially declaring biotechnology to be an environmental hazard by linguistic fiat.

To summarize, there are both substantive and rhetorical issues to consider when reviewing the precautionary principle critique of biotechnology. Substantively, figuring out how much precaution to observe is an endemic problem in risk assessment, and one that will always involve judgment. Substantive debates require that one address the nature of the hazard in question as well as the chance that it will actually materialize. Neither question can be settled by consulting the scientific evidence alone. Science cannot *settle* these debates, but it is certainly relevant and can often set boundaries for disagreement. The considerations that come up in reviewing when a biotechnology would create a hazard and then assessing the probability of actual impact are not conceptually different from those involved in reviewing any agricultural technology. But *debating* whether adequate precaution has been taken requires getting into the gory details.

Much of the public debate over precaution and biotechnology ignored the details and focused on whether US regulators, university scientists, and biotechnology companies were "taking a precautionary approach." The experience with sweeteners and Alar made this group leery of the impact that nonscientific perspectives can have on decision making. The criticisms being voiced in terms of "taking a precautionary approach" were starting to sound all too familiar to them. This community of scientists interpreted the critique as a way to push regulators into making decisions about biotechnology that would be radically inconsistent (and more burdensome) than the decisions they were making about other food or agricultural technologies. This is not to say that the biotechnology industry wasn't influenced by their economic interests, and companies making toxic chemical products and energy technologies (close relatives of tobacco or asbestos) were quite happy to jump on a bandwagon against precaution. Rationally debatable (but technically complicated) issues about whether enough precaution was being taken became obscured by the scientists' reluctance to endorse the words *precautionary approach*. This reluctance *itself* was taken to be evidence that the science community was not exercising precaution.

The Case against Agricultural Biotechnology: Social Justice

Sometimes those who appeal to a precautionary approach are calling for a broadening of the terms in which new technologies are evaluated. Here, the focus actually has less to do with uncertain risks and more to do with the inclusion of considerations that arise in connection with the four remaining categories of ethical concern. It is to these categories that we now turn.[20] Arguments stressing social justice build upon four prior critiques of agricultural research. First, the Green Revolution was criticized for tending to benefit relatively better-off (though still poor by Western standards) farmers at the expense of poorer ones. Green Revolution seeds were often distributed freely in the early years, though in some places they soon became available primarily through seed companies (who expected to be paid). In any case, the effectiveness of Green Revolution seeds depended on their responsiveness to fertilizer, and the ability to use fertilizer often correlates with ownership of draft animals, carts, or even small tractors. Actually getting benefits to the worst-off farmers often requires a heroic effort and significantly increased costs. So benefits from Green Revolution-style efforts are skewed toward farmers who were comparatively better off to start with.

Second, it is possible to broaden the point of this "skewed benefits" critique. The "technological treadmill" (discussed in Chapter 4) suggests that yield-enhancing agricultural technologies generally produce temporary benefits for early adopters at the expense of late adopters, who can eventually lose their farms entirely. Productivity-enhancing technology fuels the trend toward fewer and larger farms.[21] When we combine the differential benefit critique with the treadmill effect we find that agricultural research creates a deep tension with the goals of distributive justice. Not only do those who are comparatively better off benefit more, but those who lose their farms through the treadmill effect are thrown into more dire poverty. They join the ranks of landless labor or migrate to cities as homeless refugees. On the other side, both arguments neglect the benefits that increased yields have for consumers when there are adjustments in the price of food.

A closely related third type of argument is grounded in what development specialists call "dependency theory." In general, economic growth in developing countries has taken place by transferring technology from developed-world contexts. Whatever benefits people in the developing countries got from using the technology, the strategy put them in a position of continually depending on the original sources for continuing expertise, inputs, repairs, spare parts, and the next generation of technological advance. Farmers in particular become dependent on a host of input suppliers: seed companies; fertilizer and pesticide manufacturers; and eventually machinery producers, banks, and financial specialists.[22] The rebuttals to this critique note that developed-world farmers exhibit a very similar kind of dependency, and it is not typically regarded as a moral problem. More generally, developing-country technological dependency looks less convincing in a world where specialists from India, China, and other Asian countries are very competitive in information technology.

The fourth critique accuses developed-world agricultural researchers of "biopiracy," when they collect germ plasm cultivated by poor farmers and use it to develop certified or patented crop varieties. Here, the claim is that developed-country "seed hunters" are effectively stealing the knowledge and effort of many generations of marginal farmers who improved their local seeds through trial and error. Developed-world scientists often obtained seeds literally for pennies, sometimes simply buying them in local markets where others were shopping for food. They then brought them back to universities or company-owned laboratories where the germ plasm would be incorporated into "improved" varieties that would be sold to farmers. In some cases, they were sold back to the very farmers from whom the original germ plasm had been obtained.[23]

All four arguments were extended to agricultural biotechnology in part simply because it was, in the 1980s, the latest thing in agricultural research. Vandana Shiva was a prominent opponent of GM technology and built her case by restating (always without attribution) the litany of critiques that had been levied against the Green Revolution since the mid-1970s.[24] The biopiracy argument was particularly pertinent in virtue of the way that recombinant DNA techniques for isolating, identifying, and transferring genes introduced new ways in which intellectual property rights (IPRs) could be claimed on genetic resources. Prior to the advent of this technology, IPRs could be applied to crop varieties, and the most common form of legal protection under the US Plant Variety Protection Act recognized farmers' rights to save seed for future use. With biotechnology came patents on specific genes, multiplying the ways in which IPRs could be claimed and potentially limiting farmers' rights to save seed and interbreed purchased seed with local varieties.[25]

All these arguments from social justice make important moral points about the design and implementation of agricultural research that is intended to help the poor. However, it is important to see that when they are advanced as reasons to be concerned about biotechnology, they are inviting a fallacious inference. What they establish are conditions that *any* agricultural technology would need to meet in order to attain legitimacy. They do not single out features that make biotechnology different from most other science-based agricultural technologies, and as such do not form a basis for unilateral arguments against any and every use of biotechnology. The critiques are based on previous episodes that *did not* involve agricultural biotechnology. While Shiva may have appeared to be exposing the dark underside of an uncertain new technology, all of these criticisms were well known within the community of scientists hoping to both extend the Green Revolution and to learn from the mistakes of first-generation crop varieties. If one thinks that these are reasons to oppose biotechnology while allowing other types of industrial agricultural production technologies to move forward, one is making a grievous mistake.

The Case against Agricultural Biotechnology: Naturalness and "Choice"

A third class of arguments concerns whether GM crops are unnatural. It is clear that some people feel that they are. Prince Charles's well-known statement about genetically engineered food crops suggests that the technology is unnatural because it violates divine intentions.[26] One could certainly debate the use of genetic technologies on theological grounds, but doing so in any serious way would require consultation of specific religious traditions in a manner that goes well beyond what can be undertaken in an introduction to food ethics. When mainstream religious organizations have expressed reservations about GM crops, they more typically cite nontheological concerns relating to risk or justice. Another type of argument stresses the simple repugnance that many feel in response to genetically engineered foods. These arguments are definitely the minority view among academically trained philoso-phers, however, who point out that our ideas of what is and what is not natural change remarkable over time.[27] Critics of the view that biotechnology is "unnatural" argue that it is difficult to maintain any clear conception of "naturalness" that can be both supported by scientific conceptions of nature and yield clean ethical princi-ples for thinking one way about GM crops while thinking differ-ently about the products of traditional plant breeding.[28]

I myself have argued (and here I think many philosophers would agree) that even if concerns about the naturalness of GM crops do not provide convincing arguments for social policies that would ban or discriminate against them, such concerns do indeed provide individuals all the reason they need to avoid them as matters of personal practice.[29] Put another way, one should not be required to produce a risk assessment to justify one's preference for acting on religious or aesthetic values, beliefs about what is or is not natural, or even idiosyncratic views on what constitutes wholesome food. Even if we do not agree with these judgments, we should respect an individual's right to make dietary choices that conform to his or her personal vision of what is natural or, more generally, *appro-priate* to eat. This point leads to the fourth class of arguments in which the ethical concerns turn upon individual consumers' per-sonal values that are incompatible with eating foods from GMOs. Mainstreaming GM crops into commodity production and pro-cessing could have the effect of making it impossible for people to act on the basis of such values when consuming food. The ethical significance of this possibility is to compromise individuals' abil-ity to lead lives that conform to freely chosen religious, political, and personal values. The fact that foods enter one's body and are traditional carriers of cultural and religious tradition suggests that such compromise is a significant challenge to personal autonomy.

The policy implications point toward discussions of labeling, costs of segregating GMOs from non-GMOs, and the distribu-tion of costs from doing so. Ethical debate concerns the legitimacy and weight that should be given to such dietary concerns. Some commentators see the personal autonomy issues associated with labeling and consumer consent as a surrogate for a more system-atic divide in which the "pro-biotech" viewpoint tends to reduce all ethical issues to a cost-benefit calculus, treating all issues as resolvable in terms of the impact on total social utility. This way of thinking implies that greater social utility offsets compromise of personal autonomy. The issue comes to a head in a manner that reproduces the two-centuries-long philosophical debate between utilitarian and neo-Kantian moral theory. From the neo-Kantian or rights perspective, utilitarians exhibit a lack of concern for per-sonal autonomy that manifests itself as willingness to treat indi-viduals and their rights as "means" that can be sacrificed to pursue socially justifiable "ends."[30]

The Case against Agricultural Biotechnology: Virtue Ethics

While neo-Kantians and utilitarians debate the foundations of morality, other philosophers argue that even though we do some-times find it necessary to sacrifice individuals and their rights for more compelling social ends, the problem with the utilitarians lies in the way that they seem wholly untroubled by this kind of sacrifice. Optimizing arguments make it so easy to override individual rights that we begin to question the moral character of people who rely on them too readily (or exclusively). Shouldn't one at least regard the sacrifice of autonomy or rights as tragic and regrettable? This kind of argument spills over into the final category, aretaic objections to GMOs. The word *aretaic* is from the Greek *arete*, meaning excellence or virtue. The thrust of these concerns is to suggest either that the use of rDNA technologies is contrary to virtue, or that those who have developed and promoted GMOs have engaged in behavior that is contrary to virtue. In the latter case, lack of virtue is some-times associated with the "reductionism" of those who develop and promote biotechnology. Reductionism may refer to philosophies of science that interpret life processes as wholly reducible to physics and chemistry; to worldviews or practices that seem to regard life, nature, and even other people as lacking any spiritual dimension or sanctity; or to the belief that the subjectivity of values makes discussion of them a waste of time.[31] Evidence for poor character might also be seen in unrelenting pursuit of personal gain at the expense of others' rights or ideals of the public good, or in a tendency to misrepresent opponents and to treat their objections simply as obstacles to be set aside through whatever means. Neglect of social justice and consumer autonomy might well be interpreted as a sign of weak moral character.[32]

For many critics who advance aretaic criticisms, the weak moral character of the pro-biotech camp provides a reason to be especially cautious in dealing with them. If those who develop and promote GMOs are not to be trusted because they have poor moral character, then it is rational to be wary of these products, to see them as risky. A mutually reinforcing feedback loop begins to develop: lack of attention to key ethical issues is seen as evidence of poor moral character, and poor moral character is seen as evidence for risk (see figure 5). This evidence does not derive from facts about GMOs or their fate in the environment or the human body, but from facts about the

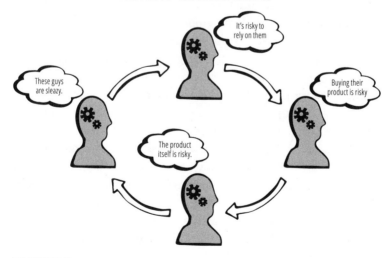

THE VIRTUE–RISK FEEDBACK LOOP

FIGURE 5

danger that we associate with people who fail to treat others with respect, or who displace serious moral issues with strategic or manipulative argumentation. As this loop becomes established, the precautionary principle can now be applied to the "uncertain risks" associated with GMOs in virtue of their shady associations. Such risks do not become better known by producing a conventional risk assessment. They can only be addressed when advocates for the technology desist from conduct that is seen as contrary to moral excellence and reestablish a basis for trust. This feedback loop allows a form of translation to cut across the argument forms, so that moral concerns about justice, choice, naturalness, and virtue are interpreted as risks, and the failure to address risks is in turn interpreted as a moral problem giving rise to still new questions about virtue. I have long argued that this feedback loop lies at the heart of much public resistance to GMOs, and that it explains the unpredictability, self-righteousness, and explosiveness of opponents' behavior.[33] I also believe that although I do not find myself to be tempted by the translations that generate this feedback, it is perfectly rational for someone with less "insider" access to the agricultural sciences to react this way.

It would be presumptuous to suggest that every critique mounted against crop biotechnology can be fitted into one of these categories, but they do encompass a large swath of the territory. Further twists can be given to these arguments by examining them in light of the growing influence of multinational corporations in the food system, or the emergence of "neoliberal" institutions such as the World Trade Organization (WTO). As with many arguments reviewed in each category, the science-based techniques of rDNA-based gene transfer are seen as problematic not so much for what they are in themselves as for the emerging shape of the global food system and some pervasive characteristics of agriculture itself. (Readers interested in pursuing the debate over biotechnology further should start with the footnotes to this chapter. They will not find themselves with a shortage of material to read.)

Connecting the Dots

Given a strong presumptive argument favoring agricultural bio-technology on the grounds that it can play a role in addressing world hunger, we can now ask a series of closely related questions. Do the arguments against it provide any basis for overturning that presumption? Are the ethical concerns about agricultural bio-technology overridden or countered by biotechnology's capacity to address world hunger? Do people who advance these concerns have a morally based reason to stifle their qualms and accept agri-cultural biotechnology because of its potential to address world hunger? Does a positive evaluation of agricultural biotechnology's potential to address hunger entail a rejection of the arguments against it? While each way of framing the question emphasizes a different set of considerations, the substantive issues raised by all of them can be addressed by working systematically through each of the five main anti-GMO arguments and examining how biotech-nology's potential to address global hunger and malnutrition pro-vides a response to them.

Biotechnology's potential to relieve world hunger provides a compelling response to the strongest interpretations of the pre-cautionary principle. Any improvement in the lot of hungry people becomes persuasive in light of Gary Comstock's demonstration of the way that broad interpretations of the precautionary approach produce self-contradictory policy prescriptions in the domain of agriculture and food.[34] If one accepts the empirical assumptions of the Borlaug hypothesis (i.e., that biotechnology *is* an important weapon against hunger), it is difficult to see how totally speculative concerns arising from uncertainty could outweigh its ethical force. On the other hand, the point of mentioning uncertainties might be to question whether biotechnology introduces novel hazards or whether the hazards typically associated with agriculture are more likely to be realized. Then we would be in a substantive debate about risk that requires more background knowledge and techni-cal expertise than can be expected in an introductory treatment.

At least for a few more paragraphs, then, let's just assume that risk assessments have been competently done and move on to social justice.

One might think that arguments from social justice would provide the most potent source of opposition to the Borlaug hypothesis. On the contrary, however, arguments from social justice stipulate a series of norms to which *any* socially just form of agricultural technology must conform. As such, these arguments spell out in more detail some of the constraints that were originally described as "widely accepted." They only apply them more specifically to agricultural technology. These arguments *do not* provide a basis for unilateral opposition to GM crops or foods, but do so only to the extent that these technologies are implemented in an autocratic manner that sacrifices the interests of many people that development assistance policies are intended to help. Concerns arising from social justice are not only compatible with the Borlaug hypothesis but represent precisely the conditions under which any ethically acceptable interpretation of the hypothesis would have to be implemented. Thus, social justice does not so much represent an ethical or philosophical challenge to agricultural biotechnology as it does a set of criteria that biotechnology must meet *in fact*. Whether any given project attempting to use biotechnology to aid the poor *does* meet these tests is an important, indeed vital, question. Yet in addressing such a question we will have effectively accepted the Borlaug hypothesis and will be working on how we should implement a given technology in a particular case.

Objections raising questions about biotechnology's naturalness or about its consistency with personal autonomy can also be dispensed with fairly quickly, if not altogether cleanly. Clearly, someone who feels that GM crops are unnatural, irreligious, or repugnant may oppose their use to aid the poor. Yet if one is inclined to regard these concerns as relevant largely to the extent that they represent legitimate personal values that deserve protection, the key issue will be whether people holding these values are given adequate opportunity to act upon them. This question points toward two distinct ethical problem sets. One concerns relatively wealthy people who purchase food in industrial food systems. Does their insistence on labels, segregation of GM and non-GM grain, and the like impose an ethically unacceptable burden on the poor? This is, of course, a more specific form of the general question this chapter explores. At this juncture in the analysis, it is fair to say that if the repercussions of insisting on one's values include the starvation and malnourishment of others, the answer must certainly be yes. Although it is important for liberal societies to give their citizens wide latitude for adopting and living out life values, and although life values relating to food may be a particularly significant subset of those life values that are protected by liberty of conscience, the evil being endured by the hungry is greater still. Protecting the religious and personal liberties of one group does not justify action or policy that causes others to starve. Of course, simply asserting that policies protecting liberty of conscience cause such dire harms does not make it so. Following this line of questioning to the bitter end would require some hard debate about the nature of both international and localized commodity markets and their capacity to deliver non-GM crops to those that want them without unduly harming the poor.

The second problem set concerns poor people who are being helped through the development of these GM crops. Have they been given adequate opportunity to apply their own values in deciding whether to adopt or eat GM crops? While it is immanently plausible to believe that concerns about "naturalness" or "repugnance" mean little to hungry people, to simply assume that this is the case fails to treat the recipients of aid with the respect they deserve. As such, there is a genuine need to introduce GM crops intended to benefit the hungry in a manner that both elicits relevant values and gives the intended beneficiaries an opportunity to accept or reject the largess of the international agricultural research system. This is not a simple task, to be sure, for it must be done in a manner that does not in itself cause suspicion about the safety of biotechnology or the intentions of donors. It is reasonable to suspect that there are unresolved ethical issues lurking here. Technically trained experts in biotechnology often express the view that their products will be eagerly adopted by intended beneficiaries, while there is strong evidence to the contrary in the form of local resistance. However, even this initial description of the task implies that the argument has shifted strongly in the direction of procedural norms that developers of GM crops must follow, much as with social justice. The problems arising in connection with intended beneficiaries' views on the naturalness of biotechnology, as well as with respect to the autonomy with which they are able to express and act on their own values, do not contradict the Borlaug hypothesis. Instead, like the arguments concerning social justice, they represent side constraints that apply to all applications of agricultural research, including biotechnology, that are intended to feed the hungry.

To summarize thus far, I have defended the Borlaug hypothesis against those who express ethical concerns in connection with the precautionary principle or with the right of relative wealthy people to make food choices that conform to their cultural values. However, I have noted that in both cases my defense depends upon resolving empirical questions in a manner that favors the optimistic assumptions of the Borlaug hypothesis. I have argued that concerns for social justice and for the autonomy of intended beneficiaries of GM crops are better interpreted as "side constraints" that do not overturn or override the Borlaug hypothesis's commitment to biotechnology but instead limit the set of ethically acceptable strategies for implementing any agricultural research program.

Virtue Ethics and the Probability of Success

What, then, are we to say about the last group of ethical concerns? Does the Borlaug hypothesis provide a reason to overlook or sublimate concerns about the moral character of those who advocate GM crops? Anyone who is inclined to think about the ethics of world hunger in outcome-oriented terms is likely to answer this question in the affirmative. If getting the hungry fed is what matters at the end of the day, then it is difficult to see why weak moral character in the people doing the feeding should contravene an otherwise successful effort. This kind of reasoning is especially relevant to a number of questions involving charitable assistance or aid. Suppose someone advocates the giving of aid not because of any feeling of moral responsibility or desire to help the needy, but because they want to be admired by others in their circle of friends, or because they want a tax deduction, or because they work for a company that will benefit economically from the aid program. These are all cases in which the person advocating assistance acts from less than virtuous motives, yet in none of these cases would the defective moral character of the advocate provide a powerful argument against the aid program. What matters here is whether the aid program can be justified on its own merits, and if the moral case for extending aid has already been made (as is the case with the Borlaug hypothesis), this justification turns upon the probability that the desired outcome will actually occur.

Any estimate of this probability must be based upon the available evidence and will involve a number of wholly empirical questions. Yet some of the important questions are *not* wholly empirical concerns, and these are questions concerning which evidence to consider and to whom that evidence is available. People such as Borlaug himself have a lifetime of experience developing new crops to address hunger and will base their estimate of the probability of success on that experience. In addition, there are technical studies on the effectiveness of past agricultural research in addressing hunger, and these studies can be extrapolated to the case of GM crops. This extrapolation is itself a technical exercise, however, and requires a firm understanding of the theory and data on which these studies rely as well as an ability to determine whether limitations in the studies affect whether they are a good basis for estimating the probability that GM crops will have similar success. In either case, then, there are elements of personal judgment that cannot be eliminated from the probability assessment. Furthermore, people who are in a position to exercise this kind of judgment are almost certain to be personally involved in agricultural research or development assistance at some level. I will call such people "insiders."

What kind of evidence is available to "outsiders," to people for whom the above-mentioned types of evidence are distinctly unavailable? Basically, their evidence that GM crops will help the hungry takes the form of insider testimony. How would a rational individual evaluate this kind of evidence? The question that rational people will ask themselves is, "Should I believe what the insiders say?" Here, the moral character of the insiders is relevant. Lacking any independent knowledge of whether agricultural research—especially research involving GM crops—will or will not help the hungry, whether it's risky, or whether it will be implemented with due consideration to social justice, a rational person will have to rely on whatever evidence *is* available to make a judgment. To the extent that it is possible to discern the motives, interests, and character of the insiders, it is entirely permissible to take this evidence into account. The Borlaug hypothesis differs from many ordinary cases in which we are inclined to ignore the motives or character of those who advocate for aid because unless we are ourselves insiders, the primary judgment we are making concerns whether or not to believe what the insiders say.

Since this book is an introduction to food ethics, many, if not most, readers are outsiders. For you, the evidence that GM crops will help address hunger consists wholly of insider testimony (for example, the articles and interviews in which Borlaug or others advocate biotechnology) and journalists' accounts, which are also based on insider testimony. If you as an outsider believe that the insiders who wrote these reports are knowledgeable, forthright, and reasonably well intentioned, you are likely to believe what the reports say. Seeing that insiders possess PhDs and professorships may be the only evidence that you have about the state of their knowledge, but as an outsider you can still bring a wealth of information and experience to bear on whether they are forthright and well intentioned. Some of the key issues are fairly obvious. Are insiders disinterested advocates, or do they stand to gain either financially or in prestige if biotechnology is pursued? How are insiders linked to the biotechnology industry? Finding out that insiders' work is often funded by industry, that their universities are seeking patents for which some them will reap financial rewards, or that private firms have a contractual right to commercialize public sector research might provide a reason to question the motives of insiders.

But some less obvious issues may be more decisive. Have insiders paid careful attention to matters that have been described above as side constraints? Have they been attentive to social justice? Have they taken pains to ensure that the intended beneficiaries of their research are truly participating in a fully informed and fully empowered way? Are they attentive to ethical, legal, and cultural concerns when they advocate for biotechnology? Given the fact that there are conflicting points of view being expressed with regard to these issues, an outsider might consider whether the insiders are dealing with this controversy in a thorough and respectful manner. That is, do the advocates of GM crops show evidence of having listened to the arguments of their detractors? Do they make responses that are on point and that either rebut their opponents' claims or explain why they are not relevant? Alternatively, do they show little evidence of having taken their opponents seriously? Do they either ignore the arguments or caricature and distort them in a manner that misses the point, making the concern seem silly? In short, are insiders committed to a serious discussion and resolution of contested issues, or do they deal with them as strategic obstacles to be overcome by whatever means necessary? Questions like this lie at the heart of aretaic concerns.

There are two points to observe. First, although talking about virtue initially suggests that we are referring to the moral character of individual scientists, this list of questions suggests that aretaic concerns also relate to the way individuals are organized into teams, or the way that activities by one firm relate to others. To cite one example, some critiques of GM cotton in India claim that farmers have been bilked into paying for low quality seeds. The evidence suggests that this is true. India has a major problem with counterfeit seeds (i.e., seeds that are labeled and sold as something that they are not). Introducing a high-yielding variety of any kind (GM or not) into this environment presents the counterfeiters with a new opportunity to practice their particular deception. But how are we to think of this from an ethical perspective? A strongly individualist interpretation would hold that the biotechnology companies can't be blamed for what the counterfeiters are doing. Indeed, they think of themselves as among the victims because they are losing money from lost sales of their product. But we might also say that there is a community responsibility here. Companies and public labs must share some of the responsibility for the poor farmers who were victimized, because it was *their* new seeds that created the opportunity for the counterfeiters. To the extent that profit motives and the prospects for career advancement through having one's technology implemented create incentives for scientists to ignore these collective responsibilities we can say that the agricultural research establishment exhibits signs of institutional corruption.

Second, given the feedback loop we discussed above, there is an important sense in which evidence about the lack of virtue in the research establishment provides some of the most convincing reasons to think that there are real, objective concerns about whether GM crops will actually help the poor. This is especially the case when institutional corruption lies at the heart of failures to work within the moral constraints. If the funders of agricultural research continuously fail to support the crucial social implementation efforts, that suggests a type of institutional corruption. If scientists pass their technology on and simply *assume* that someone else will look after the justice concerns, that qualifies as institutional corruption. If companies fail to support the public policies that would help ensure social justice, then that too is institutional corruption. In each case, a narrowly individualistic understanding of ethics will miss the point. It's not that these scientists are evil people. The problem is that a laser-beam focus on scientific questions makes them unresponsive to the larger institutional situation in which their scientific work is embedded. The corruption is magnified when individuals participate in bribery or lobby against such reforms because they will reduce short-term profits, but even those who avoid such pitfalls may be allowing less obvious problems to persist because of a head-in-the-sand attitude. Hence we cannot conclude that the Borlaug hypothesis *overrides* aretaic concerns. Rather, questions about the virtue of insiders are critical to the outsider's assessment of the probability that biotechnology will actually help the poor.

The Ethical Bottom Line

So far I have argued that despite the initial plausibility of the claim that duties to aid the hungry through agricultural research override the concerns of biotechnology's detractors, some of the detractors' concerns do act as side constraints on the way agricultural research must be implemented if it is satisfy ethical norms. Other concerns, specifically aretaic concerns, actually provide evidence that biotechnology *will not* help the poor, at least insofar as we are limited to the kind of publicly available evidence that outsiders are going to have. The nature of this evidence is positional, however. Those inside the agricultural research/development assistance establishment may have access to other evidence suggesting just the opposite. At this point it is important to recognize that the Borlaug hypothesis is not directed to insiders. The intended audience is not other agricultural researchers but thoughtful people who might be persuaded to moderate their qualms about biotechnology in light of its capacity to feed the poor.

If the issue is to be decided on the basis of publicly available information, the virtue of insiders itself becomes a relevant datum. Because I consider myself to be an insider, I find myself in an ironic position. Although I personally believe that biotechnology could be very useful in meeting the needs of the hungry irrespective of the researchers' virtue, I agree that it would be irrational for an outsider (i.e., a member of the general public) to neglect virtue arguments in assessing the ethical case for and against GM crops. But more to the point for readers, what can we say about the personal virtue of biotech insiders? Is this just a case where the public is making a tragic mistake? Or have research insiders failed to perform many of the duties that would make a fair-minded outsider see them as forthright and well intentioned? And what about the institutional organization of agricultural research and development work? Are labs, companies, and institutes organized to incentivize satisfaction of side constraints, or does social justice tend to fall through the cracks? These are the kinds of question that will help us shape our reading of aretaic concerns.

As with most cases in which one attempts to assess the virtue of a group, the record is mixed. Many agricultural scientists are thoughtful, have been attentive to issues, and have done useful service to the public's understanding of the debate. Many of the teams working to develop GM crops for poor farmers have included social scientists with a strong understanding of hunger and the institutions needed to address it. Ethical issues are discussed in many agricultural science courses, scientists have participated in public forums, and there are a number of publications by biotechnology insiders that at least attempt to weigh the pros and cons in a deliberative fashion. On the other hand, we insiders know that there are also many researchers who are either dismissive of or too busy to pay much attention to the debate. There are also a few who seem to act on truly disreputable motives. They are in it for money, the glory that comes with scientific prestige, or both.

My personal anecdotal assessment is that the truly virtuous are roughly offset by the truly disreputable, leaving the field to the dismissive and busy. This tips the balance toward a less than favorable assessment of insiders' virtue when they are viewed as a group. Although there are examples of strong efforts to secure the social justice side constraints presupposed by the Borlaug hypothesis, there are also plenty of cases where this work is either given lip service or is badly underfunded. Even among those who take up the pen and write in favor of biotechnology, very few display any evidence of having read their opponents' views. As such, I conclude that if there is blame to be distributed for the hostility that benevolent applications of agricultural biotechnology now face, a large share of that blame must be shouldered by the agricultural research community itself. An outside observer of this debate who voices their suspicions about GM crops is not expressing an unjustified skepticism.

So Where Do We Go from Here?

My conclusion is *not* an antibiotechnology conclusion. Recombinant techniques for developing new crops should be deployed in the fight against hunger, and ordinary citizens not only should support this deployment but should also seek ways to ensure that the industrial world's taste for non-GM crops does not preclude the use of biotechnology to help the hungry. As someone having extensive face-to-face contact with plant scientists, molecular geneticists, entomologists, and other agricultural scientists who are working to develop these crops, I see a number of groups and individuals whose efforts are to be applauded. Their work should be supported even as we continue to insist that sincere attempts to satisfy the conditions of social justice are expanded and better funded. Given current realities, there are very likely to be GM crops that violate one or more of the social justice side constraints mentioned above. But from an insider's perspective, we have to evaluate these efforts on a case-by-case basis. The better efforts do not create a blanket endorsement, and the problem cases do not warrant a universal condemnation.

It is, however, very reasonable for outsiders to be skeptical of this assessment. Thus these insiders should be more active in the public deliberation on biotechnology, and more respectful of opponents' point of view. This is not to say that they must agree, but they ought, at a minimum, to be capable of restating the positions of their opponents in a manner that accurately characterizes the key points. Telling the doubtful to pipe down because we are busy helping the poor is *not* a respectful response. Insiders should be willing to state clearly *why* they do not accept an argument, and they should be willing to listen carefully to any further reply that opponents care to make, replying themselves once again, if necessary. That is what virtue in the realm of public discourse demands.

In conclusion, the Borlaug hypothesis fails. Given the mixed record of agricultural insiders' willingness to ensure that their projects work within the side constraints of social justice, not to mention their failure to engage thoughtful and serious criticisms with equally thoughtful responses, one should not moderate one's qualms about biotechnology simply because Norman Borlaug, Paul

Thompson, or any other individual in the agricultural research establishment says that it will help address world hunger. In the next chapter I continue to assert that GM crops are valid tools for addressing world hunger, but I do not assert that this is a sufficient reason to stifle one's doubts, to silence one's questions, or to end one's political opposition to them.

Chapter 8

Once More, This Time with Feeling

Ethics, Risk, and the Future of Food

The fact that there has been an enormous international controversy over the use of genetic technology in food will hardly escape the notice of any moderately informed reader (see Chapter 7 for the overview). The complexity of the controversy is less obvious. On the surface it is a dispute about risks: the chance of adverse outcomes in the domain of food safety and the environment. But biotechnologies are also the latest stage in a sequence of food production technologies that include mechanical harvesters and motorized farm machinery, as well as chemical preservatives, additives, fertilizers, and pesticides. This succession of technologies has been tied to massive changes in the social organization of food production and the loss of "food literacy"—a general knowledge of where food comes from and how it is produced—among urban populations. Each of these technologies has had its own risks, though the decline in food literacy has cultivated a general ignorance about these risks and made the average person more vulnerable to a number of well-known fallacies that are tied to the evaluation of probabilities and the comparison of uncertain outcomes.

As we have seen in Chapter 7, many advocates for new food and farming technology make the case for it in terms of an imperative to "feed the world." They argue that this imperative overrides whatever qualms better-off people might feel about the brave new food. In this chapter we explore the ethics of risk and rationality more directly. Advocates for novel food technology often stress the role of scientific knowledge in addressing food risks rationally. They may implicitly assume that scientific methods produce a form of knowledge that is uniformly more precise, more reliable, and more "true" than untutored perception or experience. Given such an assumption, the ethical responsibility to think and act rationally requires one to attend closely to what science tells us is the case. Beginning around 1600 and continuing well into the twentieth century, scientists and philosophers alike developed rigorous standards for separating that which is truly *known* from mere opinion, acquaintance, anecdote, or feeling. On this view, knowledge is confined to matters of fact, and facts can be ascertained by procedures that give uniform, repeatable results in similar situations. Becoming competent in the attainment of knowledge requires cognitive discipline that generally involves quantitative skills, logic, and background familiarity with the methods and findings of other competent judges. Rationality was itself equated with such forms of competency and education.

While modern science has yielded impressive results, this approach to knowledge and rationality also became deeply implicated in repressive projects that both created and perpetuated unjustifiable patterns of exclusion, oppression, and injustice. It was almost always white property-owning men who found themselves exercising the privileges that knowledge and rationality conferred upon their possessors, while women and people of color found themselves excluded from the professions, offices, and positions where rational decisions get made. When oppressed people would complain about their plight, the standards for scientific objectivity seemed to supply the basis for dismissing their testimony on the ground that it was colored by emotion, not universalizable, unsupported by the existing science, or insufficiently confirmed by statistically significant data. Women and minorities were deemed unable to possess or transmit rational knowledge. On this basis, the knowledge elites cultivated a select and specific *ignorance* of precisely those factors that would have revealed how science and epistemology were becoming entangled in patterns of injustice. The feminist critique of traditional epistemology further argued for extending privilege to the standpoint and testimony of oppressed or marginalized people as a means for correcting several centuries of error and abuse.[1]

Feminist epistemology provides a philosophical basis for favoring the testimony of scientific outsiders in many cases where risks to health or the environment are concerned. Pharmaceutical corporations have, for example, conducted trials for new drugs or medical treatments in developing countries where poverty and lack of access almost certainly make it easy to recruit research subjects, yet there are numerous cases where abuse and deception have created hostility and resistance to further research as well as public health interventions.[2] These abuses follow more celebrated cases such as the infamous Tuskegee study, under which African American research subjects enrolled in a longitudinal study of syphilis were denied access to life-saving treatments. The incident eroded trust in and continues to influence attitudes toward the medical community.[3] On the environmental side, ecologists and veterinary researchers have ignored and marginalized knowledge of native peoples who had deep understanding of local ecosystems.[4]

By analogy, then, there are reasons to question the views expressed by technology advocates who claim to have "more rational" perspectives on the safety and environmental impact of technologies for the production, processing, and distribution of food.[5] Should our skepticism of "science-based" rationality go further than suggested in Chapter 7, where we concluded that food system insiders have been less than forthcoming in their willingness to engage reasonable concerns and ensure that social justice-based side constraints are being observed? Should we heed the alarmist buzz coming from food activists who assert that our food is positively *unsafe*, or that new chemicals, biotechnologies, nanotechnologies, or other innovations in the food system are generally (if not inevitably) linked to health hazards and environmental risks? Feminist epistemology indicates that we should privilege the testimony of oppressed and marginalized groups. Does this mean that when they express fear and distrust of new food technology we should believe *them* rather than the food system insiders whose estimate of risk is derived from a science perspective? Striking the right balance between these two points of view has become central to food ethics.

Ethics, Expertise, and Risk

Although there are convincing arguments for favoring the testimony of oppressed or marginalized people, there are also reasons to expect anyone (highly educated white males included) to be error prone when it comes to making judgments about risk. For example, a radiotherapy device was stolen from an abandoned hospital at Goiânia, Brazil, in 1987. The thieves broke open the device in an economically depressed neighborhood. This resulted in at least 4 deaths and 250 people receiving significant exposure to radio-activity. In addition, there were significant costs associated with decontamination. Any incident resulting in fatalities deserves to be called serious, and the Goiânia case revealed inadequacies in the long-term monitoring of radioactive material. Nevertheless, the social consequences of the radioactive exposure seem out of proportion to the health impacts. In the weeks and months after the incident, the entire Goiás region became stigmatized throughout Brazil. People canceled vacations and backed out of business deals, even though they would have been hundreds of miles from the site of contamination. Reportedly, pilots refused to fly planes with passengers from Goiânia, and cars with Goiás license plates were stoned elsewhere in Brazil. The adverse impact on tourism and economic development in and around Goiânia continued for years.[6]

One can understand why people might be susceptible to the effect of stigma in a case like this. We all have experience with contagious viruses such as the flu or common cold. Stories of ebola, antibiotic-resistant disease, and insect-borne disease such as malaria or dengue provide a rational basis for being cautious when unexplained deaths start to mount. What is more, assurances from public health experts might be readily dismissed, especially if there have been discrediting events that lead people to distrust medical (or radiological) expertise: "They say there is nothing to fear,

but why should we believe *them*?" For people who have reasons
to distrust the experts, the government, or elites of any kind, it
might seem perfectly rational to fear the potential for further con-
tagion or pollution when mysterious deaths are observed, or even
when they are alleged. Yet the actual mortality and morbidity from
the incident in Goiânia had a very different explanation. Here it
seems that attending too closely to the testimony of women and
people of color from the neighborhood where the radioactive exposure occurred (or airline pilots in other parts of Brazil) was just an
unfortunate mistake.

Anyone who is in the least persuaded by the feminist critique
of mainstream science and epistemology (and I am) will be inclined
to regard the writings of Vandana Shiva with earnest respect and
regard. Shiva is a charismatic and well-spoken woman who is a singularly effective advocate for the poor. Her advocacy for women
and small farmers (in much of the world, small farmers *are* women)
is tireless. Frankly, no one is better at the thirty-second sound bite
on behalf of the poor. Shiva wrote a doctoral dissertation on hidden variables in quantum theory and refers to herself as a scientist,
but perhaps it would be more accurate (and in this context more
honorific) to characterize her as a philosopher. All of these characteristics jibe with a feminist rationale for taking Dr. Shiva's testimony on GM crops quite seriously. But as I have hinted already, it is
easy for anyone to be wrong about risk.

Shiva is now probably the world's most prolific and effective
opponent of GM crops and modern biotechnology. A complete summation of the many arguments she has leveled against this technology would fill many pages. One major thrust of her critique involves
the claim that entry of GM seeds India precipitated a rash of crop
failures, indebtedness, and farmer suicides. This thrust is also supported with allegations that GM seeds have not been adequately
tested and thus may pose serious risks to human and environmental health.[7] I would not disagree with Shiva's campaign against the
mounting power of corporations or the pattern of policy evolution
in international institutions such as the WTO or World Bank, and
I certainly would not oppose her advocacy for the poor. But if we

take my two-sentence statement of her case against GM crops as a fair summary of some frequently stated arguments (and I believe that it *is* fair), we can see how in *making* this critique in the larger context of women's advocacy she creates a somewhat misleading picture of GM crops.

Start with the crop failures. There is no doubt that there have been occasional crop failures associated with GM seeds, but that is because there are occasional crop failures associated with *all* commercial seed varieties. Texas cotton growers experienced a massive failure of a Roundup-Ready® variety in 1998,[8] but reputable seed companies compensate farmers for such losses in the United States and India alike. Early failures have not deterred either US *or* Indian farmers from planting hundreds of thousands of acres in GM crops ever since. The fact that India has a problem with *dis*reputable seed dealers exacerbates the risk, and in Chapter 7 I support the claim that even reputable seed companies have an obligation to anticipate the fact that vulnerable farmers may well be cheated into purchasing inferior seed. But I would also claim that neglecting to inform a developed-world audience about these important details of seed industry practice creates a misleading picture of the risks associated with GM crops.

There is also a link between indebtedness and farmer suicides that is much larger than GM crops. A rash of farmer suicides swept across the Midwestern United States in the early 1980s when an era of easy credit (due in part to the 1970s savings and loan scandal) and high food prices came to a sudden close.[9] This was a good twenty years before any GM seeds were in existence. Farmer suicide is, in fact, a global phenomenon with a significant body of demographic and sociological research. Freelance journalist Ron Kloor has reviewed the extensive media coverage of farmer suicides in India, as well as the United States and Europe, and he notes Shiva's continuing role in linking Indian suicides to GM crops. But Kloor also reviews the extensive public health research on Indian farmer suicides, as well as testimony from experts on the inflexible and onerous financial policies of the Indian banking industry. Debt is indeed one cause of farmer suicides, but even here the tie to GM crops is very weak. The peer-reviewed research suggests that these suicides are due to a much more complex set of financial and social causes. He concludes:

> The need for Indian policy reforms that provide rural farmers with much better financial and social service resources seems clear enough. And when drought or floods victimize these farmers, the lack of a state-level safety net appears to drive some of them to suicide. Blaming farmer suicides on Bt cotton thus seems not only to be incorrect but also a distraction from the real causes of a tragic problem. One is left wondering what problem Vandana Shiva and other like-minded activists are actually interested in solving, since it does not seem to be the livelihoods of Indian farmers.[10]

Are New Food Technologies Adequately Tested?

What about claims that GM seeds have not been adequately tested and thus may pose serious risks to human and environmental health? We might begin with a logical point: it is impossible to deny *any* claim to the effect that one thing or another "may pose serious risks," because it is impossible to exclude the possibility of unknown unknowns. The idea here is that one cannot exclude the possibility that there are hazards one has yet to encounter, or mechanisms for realization of known hazards that have not been anticipated. There is also no doubt that unknown unknowns sometimes come back to bite us in the butt. The endocrine-disrupting effects of some synthetic chemicals are a case in point. As recently as 1970, no one had thought about the biophysical mechanism behind this class of chemicals, which mimic the shape of naturally occurring proteins that regulate key biological functions. The harmful effects of endocrine disruption were hotly disputed well into the 1990s, yet today it is a known hazard. Endocrine disruption is no longer an unknown unknown.[11]

The hitch here is that one cannot expect "adequate testing" to identify unknown unknowns. If one can *test* for a hazard, it cannot be an unknown unknown. Food safety and environmental impact tests anticipate *known* forms of toxicity, carcinogenicity, and ecological disruption. To be sure, there is an important difference between tests that demonstrate potential for harm with a high degree of statistical confidence and those that merely provide some evidence for harm. Advocates of the precautionary principle argue that *proving* a link to harm is asking too much. Regulatory agencies should adopt the lower burden of proof. When tests for known forms of toxicity or other forms of injury produce statistical data indicating a possible causal link to a given product or substance, regulatory agencies should take action to prohibit or control that product or substance. But as clear-cut and simple as this may sound at first, applying it to GM crops is not always straightforward. Here a case study may illuminate the problem. Rats fed Roundup-Ready maize (a GM crop) in a celebrated long-term feeding trial developed mammary tumors and had elevated mortality when compared to controls. Gilles-Éric Séralini and the other investigators for this study were careful not to allege a causal link between the GM crop on which the rats were fed and the tumors. They did conclude "that lower levels of complete agricultural glyphosate herbicide formulations, at concentrations well below officially set safety limits, induce severe hormone-dependent mammary, hepatic and kidney disruption," and that "other mutagenic and metabolic effects of the edible GMO cannot be excluded."[12] This "evidence of harm" was admittedly far short of a proven risk, but wasn't it enough to trigger a precautionary stance?

However, the plot thickens. The findings of Séralini's group stand in contrast to a meta-analysis of twenty-four separate long-term feeding studies ranging from ninety days to two years, half of which followed rats through multiple generations. The authors of *this* study write, "Results from all the 24 studies do not suggest any health hazards and, in general, there were no statistically significant differences within parameters observed. However, some small differences were observed, though these fell within the normal variation range of the considered parameter and thus had no biological or toxicological significance."[13] The Séralini group's paper sparked significant press coverage as well as critical letters from other scientists conducting similar types of work. It was retracted by the editor of *Food and Chemical Toxicology* in January of 2014 on the ground that the strain of Sprague Dawley rat used in their research is known to be prone to such tumors, and that a study to identify a carcinogenic link should have used at least fifty animals in both treatment and control groups, while the Séralini group used only ten rats in each group.[14] In their response to critics, Séralini and colleagues note that they were conducting a scientific study, and that it would be a mistake to interpret their research as the basis for a regulatory decision.

Even from this brief summary we can note some things that would give the regulators headaches. First and most obviously, we have "feeding wars" with one scientific group achieving results that are dramatically contrary to those of numerous other studies. Regulators would have difficulty interpreting such an anomalous result as "evidence of harm," even in a purely precautionary sense. Séralini and his colleagues have published explanations of why they think that others have not found these effects, but readers with an interest in this scientific controversy will need to follow that up on their own.[15]

It may be more important to notice a crucial detail that escaped the notice of virtually everyone who urged a precautionary approach to GM crops based on Séralini's findings. This controversy is actually about glyphosate, the herbicide that goes by the trade name of Roundup, rather than the genetic transformation that makes a plant resistant to glyphosate. Roundup is used in many applications that have nothing to do with GM crops. Indeed, Séralini's group has also conducted work on farmers who are exposed to glyphosate in the process of applying the herbicide as a spray.[16] As he and Robin Mesnage assert in another article, it is entirely reasonable to take a second look at chemicals such as glyphosate (which normally degrades rapidly in soil) whenever a new technology such as a herbicide resistance changes the conditions under which people or animals are exposed to it.[17] This would be true whether the resistance is achieved through genetic engineering or conventional breeding. As Séralini and coauthors have insisted, their research *does not* say anything to suggest that genetic modification is *itself* the basis for these tumors. It would be an error—albeit an error rather commonly made by other authors who hope to run with these results[18]—to link the process of genetic engineering to cancer. A regulator, however, will notice this seemingly subtle point. If there is a regulatory action to be taken here, it will very likely address glyphosate itself rather than GM crops.

The Séralini study is only one case, but it illustrates a key point. Although one can debate whether GM crops have been *adequately* tested, it would be completely wrong to infer that they have not been tested at all. In fact, GM food crops have been far more extensively tested than any other group of whole foods in human history. Here I'm making a distinction between a whole food—an ear of corn, a bean, a tomato—and a food additive or ingredient. The latter are much easier to test using standard toxicological measures. Whole foods may be resolved into thousands of different biochemical components, and a tomato grown in Hawaii may have a substantially different composition than a tomato of the same variety grown in France. (That, by the way, is why the French can make such extravagant claims for their wine.) But this kind of variability makes testing of whole foods into a process with many confounding variables.

The Ethics of Food Safety

More generally, it is psychologically and philosophically easy to raise doubts about the safety of food and logically very difficult to allay them. People do not normally think of eating food as something that is risky or that should be the focus of a detailed, deliberative evaluation. There are, however, quite a few hazards that can be encountered in eating, and most will not really surprise anyone. Some plants and animals are toxic: eating them will make you very sick and can potentially kill you. Some people have specific allergic reactions that can sicken or kill. Sometimes foods become adulterated with substances that don't belong in food, and these substances can be hazardous. Sometimes the adulteration is intentional (poisoning or negligence) and sometimes it is entirely accidental (salmonella or E. coli contamination and dangerous chemical residues). Industrial societies have developed regulatory regimes that limit and control many of the potential hazards that could possibly be encountered with food, but they do not eliminate these hazards, and they do not make food "safe."

Few people today have much sense that introducing a new food into the diet was once viewed as a source of hazard. Potatoes and tomatoes were once widely feared, and not without reason. Both plants have toxic relatives, and potatoes in particular can become toxic if they are allowed to turn green. In fact, it's relatively easy to create a toxic potato through ordinary plant breeding, simply by accidently activating the genes that make the green parts of a potato plant toxic in the root tubers that we normally eat. Potato breeders know this and are on the lookout for it. In fact, plant breeders do a lot of potentially hazardous things to get plants that farmers want. They have long used techniques that allow them to cross species lines (the claim that this is what is new about GMOs is just not true), and they have bombarded plants with high doses of chemicals or radiation to induce spontaneous mutations in the hope that they might be useful. Despite this, there is no regulatory oversight for ordinary plant breeding. The only reason novel plants have not killed or sickened people in the past is because we have been able to rely on the professional ethics of plant breeders. None of them want to produce a plant that makes people sick, and they have learned to watch for signs of acute toxicity.

Is there any reason to think that the new techniques associated with GMOs are different? Most plant scientists think not. Everyone concedes that something untoward is possible. Early in the application of these tools, there was an accidental introduction of the protein that causes allergic reactions to Brazil nuts into a plant. It was picked up quickly by the breeders and the project was canceled. It was *not* the regulatory safeguards, but the professional ethic of agricultural scientists that detected this potential hazard well before anyone even proposed that it enter the food system. So why are there any regulations at all? This is actually an important question, though perhaps only Henry Miller—a rather acerbic critic of regulation associated with the Hoover Institute—has been brave enough to raise it.[19] Here it must suffice to say that the major biotechnology companies were quite comfortable with the idea that their products would be subjected to regulatory oversight, and were active in developing the US regulatory approach that is jointly overseen by the FDA, the EPA, and the Animal and Plant Health Inspection Service (APHIS) of the USDA.

Could something go wrong? There is no doubt that it could, but most of the arguments for redoubling the regulatory oversight of GM crops (not to mention banning them altogether) apply equally to plants and animals that are developed using conventional breeding techniques. If the testing for biotechnology is not adequate, we can say that the testing for *all* new crops is not adequate. The philosophical dilemma here is a serious one. On the one hand, our tools are becoming more powerful both in their ability to introduce novelty into food *and* in terms of our ability to understand hazards. As noted, the endocrine disrupter effect was completely unknown and unsuspected only thirty-five years ago. There are powerful reasons to think that we should be more cautious about *everything*. On the other hand, being more cautious can be expensive. It costs money to do even the limited tests needed to get a GM crop approved, and there are very few (if any) not-for-profit organizations that can afford to do it. Prior to the biotechnology era, a large percentage of new crop varieties were developed by public universities and government experiment stations, including the international agricultural research centers focused on alleviating poverty and hunger. In a world of extreme caution, that era may well be over. If the public sector is out of business, it is not at all clear who will develop the crops that poor farmers need in response to climate change or other emerging challenges.

In summary, eating is potentially dangerous. It is no more unreasonable to purge GM foods from one's diet than it would have been to cancel that trip to Goiânia back in 1987. In either case, gossiping about one's fears with friends and neighbors (not to mention posting them on the Internet) is the very process that creates stigma. And once you've heard about mysterious deaths in Goiâs or rat tumors or toxic potatoes, maybe you'll suspect the motives or competence of the elites who are telling you not to worry. Philosophers of a certain ilk have been inclined to say that you would be *irrational* for doing that, but I would submit that there is some murky turf here that has not been well articulated in ethical terms. So let us see how two philosophers might address these issues.

The Biotechnology Controversy as a Philosophical Dispute

Sometime in the early years of the current millennium, Monsanto Company—a major biotechnology firm—established an ethics advisory board and appointed the Canadian philosopher of biology R. Paul Thompson to it. Thompson was widely known at the time for his work on evolution both as an explanatory framework in biology and also in ethics. He has only recently made contributions to the philosophy of agricultural science or agricultural ethics, specifically a book-length study of the debate over gene technology, *Agro-Technology: A Philosophical Introduction* in 2011. As improbable as it may seem, another philosophy professor named Paul Thompson already had several dozen peer-reviewed articles on agricultural biotechnology in print, as well as the book *Food Biotechnology in Ethical Perspective* (originally published in 1997), at the time Monsanto created its committee. He had also written or edited four other books on the philosophy and ethics of agricultural science and policy. He was the founding director of the Center for Biotechnology and Ethics at Texas A&M University and also served on the USDA's now defunct Agricultural Biotechnology Research Advisory Committee (ABRAC). A second edition of *his* book on biotechnology was published by Springer in 2007. This second Paul Thompson is also the author of the book you are reading right now.

The circumstance of dueling Thompsons provides a convenient if somewhat unconventional way to close this review of new biotechnologies, as well as the entire book's overview of ethical issues in the food system. As food ethics intersects a surprising range of social and ethical topics, bringing their interdependencies into view, the GM crop debate cuts across food ethics. The biotech controversy is in fact a proxy for a large number of ethical issues in the food system. In considering the philosophical dimensions of a broad-ranging controversy, it is always handy to have philosophers who are taking contrary views.[20]

We must begin by noting that both Paul Thompsons actually agree on the big issues. Agricultural biotechnology has made ethically significant improvements to the commercial monoculture systems in which it has been adopted, not only in the United States and Canada, but also in Latin America, India, and China. In practice, the two general types of GMO that have been widely adopted have increased food security, improved farmers' livelihoods, and reduced harmful environmental impacts. Both Thompsons agree that opponents of biotechnology have often built their case on exaggerated and unsubstantiated allegations of risk. Although it is clear that transgenic methods could be used in destructive or environmentally risky ways, both Thompsons agree that risk analysis provides the appropriate conceptual framework for regulation and that the logical structure of risk analysis provides a starting point for ethical analysis of outcomes that are potentially hazardous to human and animal health or to the environment. R. Paul Thompson (the Canadian who served on Monsanto's board) has a very fine discussion of environmental risks from horizontal gene flow, that is, the movement of genes from one plant species to another. We won't try to duplicate that in this book.[21]

Both Thompsons agree that agricultural technology can be dangerous and that the suite of tools being developed for genetic modification of plants and animals is no exception to this rule. Yet we are becoming more fully aware of the hazards that have already been introduced into agriculture every day. Chemical and mechanical technologies have increased food production, but they have done so at considerable cost to human and environmental health. We need to change. We have complex philosophical and technical problems to solve, and one dimension of that complexity surely resides in better understanding and managing the hazards of food production. There is every reason to believe that the status quo is unacceptable and no particular reason to think that using the new toolbox of genetic technologies will make things worse. The consequence of opposing biotechnology has generally been to continue with a status quo that we *know* is causing serious harm, and that is just crazy.

But there are two significant respects in which the two Thompsons do not agree. One relates to the theme of "rationality" with which this chapter began. It takes us back to the very beginning of the book and leads us to question how rationality and ethics are (or ought to be) connected. The other disagreement concerns the alternative to biotechnology. I see much more to recommend local, organic, or sustainable alternatives than does my namesake, and I fault him for failing to question the lines along which this debate is frequently drawn.

Where the Thompsons Disagree: Rationality and Ethics

R. Paul Thompson's book is structured with three introductory chapters covering scientific and philosophical background followed by three chapters reviewing aspects of the controversy over transgenic crops. I have no substantive disagreement with his summary discussions of genetics and the techniques for plant improvement. It is notable that in a world of rapid change, new methods for targeted modification have made many of R. Paul Thompson's comments on techniques somewhat obsolete only a few years after the book was published, but this is not a topic worth pursuing in the present context. His background discussion runs to one hundred pages—almost half the book—and includes a discussion of rationality that is intended to provide a basis for the evaluation of the arguments that follow. This background section may be thought of as geared to classroom use, but it is also the source of a significant philosophical disagreement between the two Thompsons.

Succinctly, R. Paul Thompson holds that logical consistency is the sine qua non of good ethical reasoning. Consistency is achieved in ethics by analyzing claims that express judgments of good and bad or right and wrong. Conceptual analysis reveals both the empirical content of such claims—what they are saying about states of affairs in the world, and also the more general normative principles that are implied by them. Normative principles provide general statements about what things are good or bad, and why. They also indicate decision rules for moving from an account of what is good and bad to a prescription, a statement of what should be done in a given circumstance. R. Paul Thompson leans toward consequentialism without fully committing to it, insisting only that whatever general normative principles one commits oneself to, one must be logically consistent in drawing inferences from them.

Contrarily, I believe that our lives are saturated with inconsistency. Following recent work in cognitive science (or for that matter, the thought of David Hume or John Dewey), I see ordinary decision making—including most opinion formation—as a generally unreflective process dominated by affect, habit, and the largely unconscious substitution of easy questions for the more difficult processing of quantitative thinking, making logical comparisons, and careful testing of analogies needed for deliberative problem solving. Notice that this is a potentially telling point, given what has already been said: people do not normally think of eating food as something that is risky or that should be the focus of a detailed, deliberative evaluation. I would not deny that consistency and the identification of general principles is important some of the time. It is difficult to prove a theorem in logic or do a proper risk analysis without these cognitive skills, for example. But application of these skills also presupposes a prior conceptualization of what matters about the situation at hand.

People doing risk assessments do not always have the answer when it comes to "what matters about the situation at hand." Indeed, past toxicological studies neglected to include enough women and occasionally overlooked impacts on the female anatomy. Mammary tumors such as those found on Séralini's rats would be a case in point. Although GM crops have been more thoroughly tested for known toxic hazards than any other whole food, none of the regulatory testing would ever raise questions about linkages with financial policies or seed counterfeiting. They would never consider the possibility that GM crops might elevate the risk of farmer suicides. Now I hasten to add that I'm not backing off from what was said previously. It would be misleading in the extreme to assert a unique linkage between genetic engineering and untested impacts on women, or with farmer suicides. Many elements in the food system could be connected with these hazards, not just GM crops. I'm still a philosopher and I wouldn't give up on logical consistency *entirely*. But my logical point would be to stress a general failure to think broadly enough about food.

We should admit that our unreflective judgments work for us most of the time, and also that even our most deliberative and logically careful efforts are only possible because they rely on past judgments and deliberations, many of which were made by other people who were not even remotely thinking about the problem at hand. We can apply careful logic when there is agreement about what the problem actually is, but that is manifestly *not* the case in the GMO debate. This is a debate in which the application of "careful logic" usually involves insisting that one's own diagnosis of the problem is the fixed referent for problem solving. And that diagnosis also presumes general metaphysical views on how the world works. But that looks suspiciously like a power move to people who don't agree, and power moves only make the disagreement deeper. One of the reasons why it's relatively easy for plant scientists to agree about GMOs is that they have already made a substantial cognitive investment in developing a shared framework. I am impressed by their investment, and I take it seriously, but I don't think that ethics requires people who can't tell a potato plant from Arabidopsis to bow to the consensus of plant scientists on matters of what they will eat for dinner.

Power moves also initiate the vicious cycle of doubt and distrust Chapter 7 discusses. "If these guys would treat me like that," one thinks to oneself, "shouldn't I be leery of them? And if I'm leery of them, how could I possibly have confidence in their products? Those products are probably *dangerous!*" If one has frequently been marginalized, neglected, or had one's standpoint ignored and even systematically eliminated and repressed by people who claim to be doing science, it seems entirely *reasonable* to react this way. What feminist epistemology brings to the table is a way to understand why the strict rigor and standards of proof that may be vital to certain components of a scientific inquiry should not be generalized as criteria for judging the thinking and attitudes of other human beings. Science may not necessarily require interacting with human beings who think differently, but ethics does. Living in the real world gives every person experience that is enormously rich and multitextured. Just because someone's experience can't be generalized or pigeonholed into a given conceptual framework doesn't mean that it isn't valuable. Indeed, the fact that it can't be generalized may be a key to its value.

R. Paul Thompson has taken a type of thinking that human beings only do with great effort and suggested that this should be the standard for ethics. It is, to hark back to the Introduction of *this* book, narrowly conceptualized in the domain of assimilative learning skills. It suggests that "getting it right" is the vector that proceeds out from the origin toward the upper right in our figure of the learning cycle (see figure 6). Ethics becomes solely a problem in building a theory that consistently and comprehensively organizes all relevant information. This may be great fun for philosophers sitting alone in their study (to use one of Hume's metaphors), but once we get back into the world, all of that fades away and there is more than one way to be right (or wrong). Ethics will be more usefully characterized as a way of engaging with others on matters where there are multiple ways of defining the problem, conflicting economic and political interests, no clear criteria for a solution, pervasive uncertainty, and significant costs of being wrong.

Anyone addressing GM crops must make a number of simplifying moves, but I would also argue that R. Paul Thompson's exclusive focus on the environmental risks of gene flow fails to provide an adequate portrayal of the complexity of the issues that have been raised against agricultural biotechnology. The "labeling debate," for example, is entirely missing from *Agro-Technology*. In 1997, I argued that then-extant policies foreclosed dissenters' ability to avoid consuming biotech products in an unjustifiable way.[22] The problem was partially rectified by organic labels that provide consumers with an "opt out," but the food industry has continued to insist that standards for consumer choice should always be supported by science. As indicated above, one should not have to produce a logically consistent and scientifically rational argument to defend their values for food choice. To be sure, I do not argue for *mandatory* labels. Instead, "GMO free" should be seen as a values-oriented standard much like labels that protect opportunity to follow religious dietary codes or policies that prohibit the use of cat and dog

THE LEARNING CYCLE AND THE FOUR PHASES OF INQUIRY

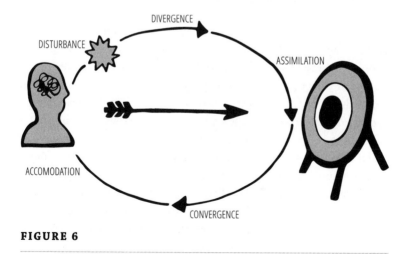

FIGURE 6

in prepared meats.[23] Such rules are proliferating in the food system, with standards such as "fair trade," "humanely produced," and "local" joining the mix over the last decade.

The retreat behind science-based labels ignores the fact that US regulatory standards are replete with value judgments. Although they generally function within a limited domain or context, in total these regulatory value commitments betray blatant logical contradictions. FDA policy holds that labels should be based on a scientifically certified link to nutrition or health. The FDA won't let a manufacturer use words like *light* on their banana cream pie unless it is significantly lower than some non-light product in fat or calories. Prior to the implementation of this rule in the 1990s, food companies might have called a cream pie "light" because its color was less deep than usual. However, the USDA insists on country-of-origin labels in the absence of any reason to think that *they* bear on the safety or environmental impact of food, and our grocery stores ban meat from dogs or cats with no supporting risk assessments whatsoever![24] And why not, say I? There is no reason why people should be required to produce a scientific risk assessment simply to defend an ethical, religious, cultural, or political food preference. Insisting otherwise would be what Ralph Waldo Emerson characterized as "foolish consistency": the hobgoblin of small minds.

Since no links to safety or nutrition have been found for GM crops, "non-GM" labels have been discouraged at FDA (though the energy devoted to enforcement of this policy has varied over time). But I would argue that people might want to seek out or avoid foods from GM crops for the same kinds of reasons that they seek out or avoid products from certain countries. Their motives may be political, religious, or just plain arbitrary, and they may involve beliefs about risk and safety. Recall the melamine scandal in Chinese powered milk production mentioned in the Introduction. People might avoid imports from China based on the same kind of reasoning that led them to cancel their trips to Goiânia. Food-safety professionals might react to this in much the same way that radiological health experts reacted to the stigmatization of the Goiás region in 1987. But this does not imply that people should be deprived of truthful (even if deemed irrelevant) information that they want to make whatever choice they take to be appropriate. I don't know that I want to generalize this rule of thumb to all areas of life, but it is surely a reasonable guidepost for thinking about food choices.

Where the Thompsons Disagree: The Philosophy of Agricultural Science

R. Paul Thompson's book concludes with a chapter reviewing the organic alternative and a brief but important chapter discussing agriculture in low- and middle-income countries. His discussion of impacts on poor countries echoes work by Robert Paarlberg (also a member of Monsanto's advisory group). As discussed in Chapter 7, Paarlberg has argued that the policies and practices that have denied African farmers the benefits of biotechnology are unconscionable. He is thinking primarily of Europe, where GMOs were banned for a time and where labeling is mandatory. European supermarkets continue to enforce storewide bans on products containing GM crops, even though EU policy now permits them to be both sold and grown. Since many African nations view Europe as their primary export market, there has been reluctance to allow GMOs in African agriculture for fear that this will tarnish their reputation for quality. Paarlberg writes that Europe is forcing Africans into an underperforming and unprofitable version of organic farming.[25]

R. Paul Thompson's *Agro-Technology: A Philosophical Introduction* characterizes conventional non-GMO monoculture as the status quo and presents biotechnology and organic agriculture as competing alternatives to it. Anyone who has spent any time as an insider in agricultural research organizations will recognize that there is a certain logic to this approach. Scientists who study or teach organic and alternative approaches in agricultural universities seldom collaborate with molecular biologists who work on transgenic methods of plant transformation. I could attest to many instances of tempers flaring and uncharitable accusations flowing in both directions. In the world of farmers, the situation is only a little different. Many organic growers are bitterly opposed to GMOs. On one particularly memorable occasion, I was angrily told that GMOs are an *alien* technology by a relatively well-known American advocate of organic farming, and his pronunciation of the word *alien* clearly conveyed the sense that they are "not of this earth." On another occasion, a Mexican peasant farmer told me that he thought I wanted to kill him and take his land, all because I had failed to condemn GM crops in a talk that I just given.

However, relatively few farmers in the United States would characterize GM crops as an alternative to non-GM monoculture. GM varieties of corn, cotton, and soybeans have been widely and rapidly adopted by conventional farmers, some of whom also maintain fields of organic crops. Bt crops *have* reduced the use of more hazardous pesticides. However, while herbicide-tolerant crops led to an initial decrease in herbicide use and the substitution of more benign glyphosate herbicides for more hazardous chemicals such as atrazine or 2,4-D, farmers have taken to spraying glyphosate so frequently that total herbicide use has actually increased. Important weeds are now becoming to resistant to glyphosate, replicating a pattern that has long been characteristic of conventional agriculture. GM crops have *not* led to a decrease of monoculture (farming that focuses on a single crop) in general, and may have encouraged some corn and cotton growers to reduce rotations of crops that were intended to control pests or restore soil fertility.

Given this, we should reject R. Paul Thompson's formulation of *two* alternatives to the status quo and recognize that whatever potential biotechnology *might* have for the future, the current GM crops are following the path of other industrial monocultures. It is possible to trace this path backward to an era when GM crops were not even the glimmer in some molecular biologist's eye. Historian Frank Uekoetter argues that agricultural science institutions unwisely shed key initiatives from their portfolio beginning in the 1930s. Uekoetter writes that the obvious effectiveness of synthetic fertilizers made their adoption attractive to farmers. Synthetic fertilizers depend upon industrial production of ammonia using the Haber-Bosch process, which extracts nitrogen from the air.

The German chemist Fritz Haber demonstrated the process in in 1910, and Carl Bosch developed large-scale production facilities to supply nitrogen to the German munitions industry by 1913. After World War I, these production facilities were repurposed to produce ammonia for synthetic fertilizer. Since the money for building these plants had come from military spending, the fertilizer could be made available to farmers relatively cheaply.

At the same time, ecologically oriented biologists in Germany had made what turned out be an unfortunate intellectual alliance with anti-Darwinist philosophies that postulated the existence of a "life force." Twentieth-century science was becoming more quantitative and reductive, and it was not long before speculating about a life force was simply not respectable in scientific circles. What was lost (or at least substantially de-emphasized) included soil biology and systems ecology, on the one hand, and an emphasis on what Liberty Hyde Bailey had (perhaps misleadingly) referred to as "permanent agriculture," on the other. The advocates of more ecologically oriented farming and more systems-oriented research found themselves out of favor in German agricultural science. I suspect something similar was happening worldwide. Uekoetter argues that the current split in agricultural science institutions reflects a continued narrowing of perspective. Succinctly, post-war agricultural science has been dominated by biophysical approaches grounded in chemistry and genetics, and by socioeconomic values that emphasize aggregate productive capacity and year-to-year farm profitability.[26]

Agricultural researchers also abandoned methods of study that identify practices of acknowledged master farmers with the aim of discovering the scientific basis for their success. Against the protests of mainstream agricultural scientists such as Sir Albert Howard, methods for agricultural science drifted away from active on-farm research. Today, researchers undertake on-farm trials almost exclusively as an evaluation process for technologies developed in dedicated agricultural laboratories and test plots. Although agricultural scientists continue to express admiration and respect for practicing farmers, there are very few research programs in which farmers are involved as active collaborators. Organic farming gathered steam in the 1960s and 1970s as a farmer-based initiative to share knowledge on soil management and integration of complex cropping and livestock production systems. It was often conjoined with a social agenda of small-farm survival and promotion of what Bailey had called "country life." Contemporary organic farmers developed their methods with very little help from institutionalized agricultural science, and there has been very little research on complex crop complementarities and rotations that characterize the most successful organic farmers, not to mention the curious effectiveness of homeopathic crop supplements developed by biodynamic producers.[27]

But there is also virtually nothing in the contemporary organic standard that reflects a significant commitment to any of the thrusts that initially gave rise to organic farming methods, and it would be misleading to suggest that the majority of farmers producing organically certified commodity crops are utilizing the most productive or sustainable methods that have been developed under the aegis of organic methods. The standards for marketing farm commodities under the USDA certified organic label limit the use of many chemical inputs but do not mandate the use of complex methods that farmers have developed to maintain soil fertility; manage genetic diversity; limit damage from insects, fungi, and disease; or ensure compatibility with wild flora and fauna. Neither do they address socioeconomic principles intended to ensure the sustainability of rural economies and the survival of small-scale farms. It is thus highly questionable that the organic designation identifies what was philosophically important about the original thrust toward alternative farm production methods.[28]

Despite important exceptions and significant progress over the last decade, public-sector agricultural science continues to reflect significant weakness in ecology, including microbial ecology and its relationship to soils and plants. What is more, sustainability continues to be understood almost exclusively in terms of global food availability, with little appreciation of the biological or social integrity of local, regional, or global food systems, much less the adumbration of commodity monoculture through the food supply chain or its potential for impact on diet and health.[29] At the same time, there has been a subtle shift away from the transgenic methods that are the centerpiece of R. Paul Thompson's characterization of agrotechnology. Genomics and marker-assisted breeding are now deployed in accordance with mainstream agricultural science's continued commitment to genetic, chemical, and mechanical technology intended to optimize biological yields and economic return on capital investment. Given this trend, Thompson's distinction between conventional agrotechnology and biotechnology is not meaningful.

R. Paul Thompson does not totally reject the organic alternative. Yet in systematically presenting biotechnology and organic agriculture as mutually exclusive alternatives to *conventional* agricultural technologies, he creates a misleading picture. There are regions (such as Europe) where chemicals that might be replaced by Bt crops continue to be used, and some chemical use could probably be reduced if transgenic approaches proven in corn, soy, and cotton were applied in wheat monocultures. But as yet no GM crop has created a significant saving in consumption of energy or water, and unlike the dwarf varieties of the Green Revolution, GM varieties have not increased the yield potential of major cereal crops. From a North American perspective, GM crops *are* conventional agriculture. This is, in fact, part of the problem. Whether organic or not, agro-ecology applies management-intensive methods deploying complementarities among plants and animals that are atypical of monoculture or that concentrate on plant, soil, and animal interactions and rural community development. There is no conceptual reason why they cannot be pursued simultaneously with genomics-based breeding or even gene transfer.[30] It is not surprising that highly capitalized for-profit input firms find little of interest in methods that rely so heavily on farming skill and management capability, but regions that still have a significant percentage of their population engaged in farming have many reasons to pay attention to them.

It is thus not helpful to characterize the philosophy of agricultural technology as a contest between agricultural biotechnology and organic agriculture. It is especially misleading when it comes to strategies for applying agricultural science in less industrially developed regions. What is more, to equate current practices of poor African farmers with an organic approach, as Paarlberg has, betrays ignorance of the emerging science and current technology of successful organic practitioners. The organic agriculture being practiced in Europe, Australia, and the United States today bears little similarity to the practices of resource-poor smallholders in Africa, Latin America, or those parts of Asia that remain underdeveloped. Not only is it technically complex, contemporary organic production is now the focus of significant research and development that is introducing new methods for composting, weed control, and greenhouse production. Advocates of both biotechnology *and* organic systems too often compare the most advanced and optimistic interpretation of their favored approach to the least successful applications of the alternative. In the case of biotechnology advocates, this generally means that unproven technologies and methods still in the development stage are described as if they were established achievements. Golden Rice (which may yet emerge as a great use for genetic modification) is an application that was plumped on the cover of *Time Magazine* a good fifteen years before it had any chance of being grown by farmers.

A Final Word

Although they are often overstated, philosophical dichotomies can be helpful in articulating how underlying assumptions or unreflective categories of thought steer human behavior along one trajectory rather than another. In this regard, there is a more helpful dichotomy for understanding contemporary agricultural science. There is a philosophy that regards agricultural plants and animals (along with the land and water resources on which they are based) as a general technological platform whose ultimate utilization should be determined by the priorities of the industrial economy.

Whether these tools and resources are allocated to the production of fuel, animal protein, starches, high-quality nutrients, or recreational use is largely a matter for markets to decide. The scientist's role is to abet this process by increasing the efficiency with which any of these goods can be brought forth. This *industrial philosophy of food* guides the research of most scientists deploying molecular methods, whether for transgenic crops or advanced breeding. Given climate change, scarcity of water, and growing awareness of problems with chemical inputs, scientists working in this philosophical framework are far more cognizant of external costs than they were thirty years ago. Without regard to whether or not they are using what R. Paul Thompson calls biotechnology, they are indeed far more dedicated to an environmentally responsible agenda than were the scientists who produced the last generation of agricultural technology—what Thompson calls "conventional agriculture." But it is not clear that there has been a fundamental shift in their philosophy of agricultural science or their understanding of food.

In this industrial philosophy, food is just one commodity among many. Industrial consumers want food, to be sure, but they also want health care, clothing, smart phones, vacation travel, video games, music, television, and a comfortable place to sleep. One claim of the liberal philosophy envisioned by John Stuart Mill is that we should leave to the invidual the decision of how to allocate one's resources to each of the enumerable goods that are available in industrial societies. Each person will choose among them based upon cultural background, religion, education, and overall view of what matters in life. The tolerance for differences in religion and culture that Mill's view recommends is one of its chief virtues. But Mill's liberalism became wedded to a view of rationality that emphasized maximization of consumption, and this pairing became institutionalized in contemporary science, government, and business. As we minimize our expenditure of time, effort, and wealth on the consumption of any one good, new opportunities for the consumption of other goods open up. And again, food is just one good among others. The recognition of environmental limits has begun to constrain the quest for endless consumption, but it has not altered the fundamental commitment to efficiencies that increase opportunities for the consumption of highly segregated and Balkanized goods.

Contrast this with a philosophy that sees agriculture not simply as a technological platform but a set of human practices and social institutions that fulfill a wide array of crucial functions, some of which may not even be fully understood or appreciated. These practices and institutions feed human beings, to be sure, but they can do so well (in the sense of encouraging nutritionally balanced diets) or poorly. Foodways also mediate a large set of human-ecosystem interactions and, when working well, provide rapid and reliable feedback on the stability and sustainability of those interactions. Furthermore, they provide a basis for institutions—such as the self-reliant household—that multiply and link to many different goods throughout society. The social functions of farming and food technology have undoubtedly become increasingly obscure as urbanization and industrialization of the food system have progressed. Yet problems such as rising obesity and the social movements toward local production or "slow food" provide evidence that the organization of tomorrow's food system may need to reflect much more than efficient production of commodities.

In *The Agrarian Vision*, I argue that throughout history we have called philosophies like this "agrarian." I recognize that this word carries a lot of baggage and has little resonance for many people brought up in radically industrialized societies. Some would associate the word *resilience* with this kind food system, and others just assume that this kind of integrated approach is implied by the word *sustainability*. Yet another perspective would tie this approach to the call for social justice. Here some evident failures in the industrial philosophy come to forefront. In leaving the allocation of resources to the individual and emphasizing efficiency, we create a system where those who *have* wealth create the demand that shapes the overall economy. In leaving the total allocation of resources to an impersonal market, we drive everyone into an economizing mentality with little room for compassion or solidarity with the people who *make* the commodities that we consume. Food is, again, just one commodity among others, so the people whose work feeds us can easily wind up being unable to feed themselves. I would inject a note of caution against simply presuming that a more integrated food system will also be more just. We have many examples of unjust but highly integrated food systems in humanity's past. Yet calls for "food sovereignty" and new ways of obtaining food that emphasize interpersonal connections—farmers' markets, co-ops, and eating local—do have the potential to open our eyes and then our hearts to the social dimensions of our food system.

Whichever side of *this* dichotomy one wishes to take, or whether one hopes (as I do) that they could engage one another in a creative, dialectical give and take, the dialogue between an industrial ethic and agrarian, integrated philosophies of food is a more helpful way to understand the debates in contemporary food ethics. I believe that engagement in this dialectic—being willing to listen, read, and try to understand those who occupy different roles and come from diverse walks of life—takes priority over policing the logical rigor of others' arguments. The basics of food ethics, like public philosophy in general, may demand more tolerance for inconsistency than analytic philosophers have become accustomed to. It may be more important to adopt a style in ethics that avoids premature closure on normative issues rather than one which rushes to judgment. It may be better to find those points where issues intersect, and then to engage with whomever one finds at that intersection, however it was they got there. This may require that philosophers learn something about whatever they are writing about, instead of just employing theory and conceptual analysis. At the same time, philosophy has a time-honored willingness to be contradicted by someone with a different view and to explore the connections and tensions implied by difference with patience and respect. That practice will serve us very well.

And on that note, it is someone else's turn to engage.

Introduction

1. "Fresh Strawberries from Washington County Farm Implicated in E. coli O157 Outbreak in NW Oregon," United States Food and Drug Administration, accessed June 6, 2014, at http://www.fda.gov/safety/recalls/ucm267667.htm.

2. Kirti Sharma and Manish Paradakar, "The Melamine Adulteration Scandal," *Food Security* 2, no. 1 (2010): 97–107.

3. Ann Vileisis, *Kitchen Literacy: How We Lost the Knowledge of Where Our Food Comes From, and Why We Need to Get It Back* (Washington, DC: Island Press, 2008); Warren Belasco, *Meals to Come: The History of the Future of Food* (Berkeley: University of California Press, 2006).

4. R. Douglas Hurt, *American Agriculture: A Brief History*, rev. ed. (West Lafayette, IN: Purdue University Press, 2002).

5. Warren Belasco, *Appetite for Change: How the Counter Culture Took On the Food Industry*, 2nd ed. (Ithaca, NY: Cornell University Press, 2007).

6. John Dewey, "The Reflex Arc Concept in Psychology," *Psychological Review* 3.4 (1896): 357–370. In this article, Dewey is arguing against the stimulus-response model in psychology and arguing for what he takes to be a more adequate interpretation of how the behavior of an organism responds to a set of environmental conditions. Dewey lost this debate with the behaviorists, but for my purposes here it is more interesting to notice this early and highly illustrative account of what he later came to associate with learning and inquiry. The model is developed in connection to ethics in his *Human Nature and Conduct* (1922) and receives its most exhaustive treatment in *Logic: A Theory of Inquiry* (1938).

7. David A. Kolb, *Experiential Learning: Experience as the Source of Learning and Development*, vol. 1 (Englewood Cliffs, NJ: Prentice-Hall, 1984).

8. David A. Kolb, Richard E. Boyatzis, and Charalampos Mainemelis, "Experiential Learning Theory: Previous Research and New Directions," *Perspectives on Thinking, Learning, and Cognitive Styles*, edited by Robert J. Sternberg and Li-Fang Zhang (Mahwah NJ: Lawrence Erlbaum Associates, 2001), 227–247.

9. Ibid.

Chapter 1

1. See, for example, Victor H. Lindlahr, *You Are What You Eat* (New York: National Nutrition Society, 1942). According to "The Phrase Finder," the earliest rendering in English appeared in a 1923 Bridgeport Telegraph advertisement for a local meat market. See http://www. phrases.org.uk/meanings/you%20are%20what%20you%20eat.html, accessed December 13, 2013. My search in Google Scholar turned up over sixty citations to scholarly articles having the phrase "You are what you eat" in the title, sometimes followed by a subtitle and in a few cases ("You are what you emit") with wordplay that references the original formulation elliptically.

2. See Hub Zwart, "A Short History of Food Ethics," *Journal of Agricultural and Environmental Ethics* 12, no. 2 (2000): 113–126.

3. Clarence J. Glacken, *Traces on the Rhodian Shore: Nature and Culture in Western Thought from Ancient Times to the End of the Eighteenth Century* (Berkeley: University of California Press, 1973).

4. Michel Foucault, *The History of Sexuality* (New York: Vintage Books, 1990).

5. Kelly Oliver, *Animal Lessons: How They Teach Us to Be Human* (New York: Columbia University Press, 2009).

6. Tristram Stuart, *The Bloodless Revolution: A Cultural History of Vegetarianism From 1600 to Modern Times* (New York: W. W. Norton, 2007).

7. Caroline Korsmeyer, *Making Sense of Taste: Food and Philosophy* (Ithaca, NY: Cornell University Press, 1999).

8. John Stuart Mill, *On Liberty* (New York: Library of Liberal Arts, 1956 [1859]).

9. Thomas Nagel, "Poverty and Food: Why Charity Is not Enough," in *Food Policy: The Responsibility of the United States in Life and Death*

Choices, ed. P. Brown and H. Shue (Boston: The Free Press, 1977), 54–62; Henry Shue, *Basic Rights: Subsistence, Affluence, and US Foreign Policy* (Princeton, NJ: Princeton University Press, 1980); Onora O'Neill, *Faces of Hunger* (London: G. Allen & Unwin, 1986); Thomas Pogge, *World Poverty and Human Rights* (Cambridge: Polity Press, 2007).

10. Peter Singer, "Famine, Affluence and Morality," *Philosophy and Public Affairs* 1 (1972): 229–243; *The Life You Can Save: How to Do Your Part to End World Poverty* (New York: Random House, 2010); Peter K. Unger, *Living High and Letting Die: Our Illusion of Innocence* (New York: Oxford University Press, 1996).

11. Slavoj Žižek, "Cultural Capitalism," https://www.youtube.com/watch?v=GRvRm19UKdA, accessed June 19, 2014. See also Slavoj Žižek, *First as Tragedy, Then as Farce* (London: Verso, 2009).

12. Stuart begins his study of English vegetarianism with the religiously motivated views of Thomas Bushell (1594–1674), Thomas Tany (1608–1659), and Roger Crab (1621–1680).

13. Tom Regan, *The Case for Animal Rights* (Berkeley: University of California Press, 1983).

14. Peter Singer, *Practical Ethics*, 2nd ed. (New York: Cambridge University Press, 1993).

15. Frances Moore Lappé, *Diet for a Small Planet* (New York: Ballantine Books, 1975).

16. Henning Steinfeld, Pierre Gerber, Tom Wassenaar, Vincent Castel, Mauricio Rosales, and Cees de Haan, *Livestock's Long Shadow: Environmental Issues and Options* (Rome: Food and Agriculture Organization, 2006). It should be noted that the authors of this report interpret it as an argument for more intensive animal production—which has lower emissions per pound of product—rather than for eliminating animal products from the diet. Anyone wishing to follow up on the climate ethics of emissions from animal production should be aware that the empirical details are not as simple as might be implied in the text. While methane is a very potent greenhouse gas, it is far less long-lived in the atmosphere than carbon dioxide. See the discussion in Chapter 6. Also: Sara E. Place and Frank M. Mitloehner, "Contemporary Environmental Issues: A Review of the Dairy Industry's Role in Climate Change and Air Quality and the Potential of Mitigation through Improved Production Efficiency," *Journal of Dairy Science* 93, no. 8 (2010): 3407–3416; "Beef Production in Balance: Considerations

for Life Cycle Analyses," *Meat Science* 92, no. 3 (2012): 179–181.

17. Simon Fairlie, *Meat: A Benign Extravagance* (White River Junction, VT: Chelsea Green Publishing, 2010).

18. Mary Midgley, "Biotechnology and Monstrosity: Why We Should Pay Attention to the 'Yuk Factor,'" *Hastings Center Report* 30, no. 5 (2000): 7–15.

19. Martin W. Bauer, John Durant, and George Gaskell, eds., *Biotechnology in the Public Sphere: A European Sourcebook* (London: NMSI Trading Ltd, 1998).

20. Paul B. Thompson, "Why Food Biotechnology Needs an Opt Out," in *Engineering the Farm: Ethical and Social Aspects of Agricultural Biotechnology*, ed. B. Bailey and M. Lappé (Washington, DC: Island Press, 2002), 27–44.

21. Anthony Giddens, *Modernity and Self-identity: Self and Society in the Late Modern Age* (Palo Alto, CA: Stanford University Press, 1991), 81.

22. Paul B. Thompson, *The Agrarian Vision: Sustainability and Environmental Ethics* (Lexington: University Press of Kentucky, 2010), 136–154. See also Albert Borgmann, *Real American Ethics: Taking Responsibility for Our Country* (Chicago: University of Chicago Press, 2006).

23. Eggs are discussed at some length in Chapter 5. The claim being made here should be qualified in light of that discussion.

24. There is a debate to be had in the philosophy of economics on this point. Bengdt Brülde has adapted an argument that Walter Sinnott-Armstrong has made with respect to carbon emissions: the contributions that any individual makes toward determining market demand are so miniscule that these purchases are causally inefficacious and hence morally meaningless. So in fact even the *purchase* makes no difference on Brülde's view. Economists themselves are divided on this point. Bailey Norwood and Jayson Lusk dispute Brülde's claim with respect to demand for animal products, for example, arguing that *every* consumer purchase of meat is reflected in market signal that is inducing animal producers to raise livestock. I tend to agree with Norwood and Lusk, and I wonder if the views of Brülde (and possibly Sinnott-Armstrong) might not be based on the same fallacy behind some of Zeno's paradoxes. In short, I think that integral calculus explains why one's purchases *do* make a difference. Nevertheless, the point that I am making in this chapter is simpler: it's not the act of

eating but the act of buying that is locus of the relevant social causality. See Bengdt Brülde and J. Sandberg, *Hur bör vi handla? Filosofiska tankar om rättvisemärkt, ve-getariskt och ekologiskt* (Stockholm: Thales, 2012); Bailey Norwood and Jayson Lusk, *Compassion by the Pound: The Economics of Farm Animal Welfare* (New York: Oxford University Press, 2012).

25. John Rawls, *Political Liberalism* (New York: Columbia University Press, 1993); Jürgen Habermas, *Moral Consciousness and Communicative Action* (Cambridge, MA: MIT Press, 1999).

26. I think there is less consistency in this usage than in Rawls and Habermas, but for a representative example see Judith Butler and Joan Wallach Scott, *Feminists Theorize the Political* (New York: Routledge, 1992).

27. Michel Foucault, *Abnormal: Lectures at the Collège de France, 1974–1975* (London: Verso, 2003), 99–104

28. Kari Marie Norgaard, Ron Reed, and Carolina Van Horn, "A Continuing Legacy: Institutional Racism, Hunger and Nutritional Justice in the Klamath," in *Cultivating Food Justice: Race, Class and Sustainability*, ed. A. H. Alkon and J. Agyeman (Cambridge, MA: MIT Press, 2011), 23–46.

29. Volkert Beekman, "You Are What You Eat: Meat, Novel Protein Foods, and Consumptive Freedom," *Journal of Agricultural and Environmental Ethics* 12, no. 2 (2000): 185–196.

Chapter 2

1. Barry Esterbrook, *Tomatoland: How Modern Industrial Agriculture Destroyed Our Most Alluring Fruit* (Kansas City, MO: Andrews McMeel Publishing, 2012).

2. Deborah Fink, *Cutting into the Meatpacking Line: Workers and Change in the Rural Midwest* (Chapel Hill: University of North Carolina Press, 1998). See also Michael S. Cartwright, Francis O. Walker, Jill N. Blocker, Mark R. Schulz, Thomas A. Arcury, Joseph G. Grzywacz, Dana Mora, Haiying Chen, Antonio J. Marín, and Sara A. Quandt, "The Prevalence of Carpal Tunnel Syndrome in Latino Poultry Processing Workers and Other Latino Manual Workers," *Journal of Occupational and Environmental Medicine* 54, no. 2 (2012): 198–201.

3. Annette Bernhardt, Ruth Milkman, Nik Theodore, Douglas Heckathorn, Mirabai Auer, James DeFilippis, Ana Luz González,

Victor Narro, Jason Perelshteyn, Diana Polson, and Michael Spiller, *Broken Laws, Unprotected Workers: Violations of Labor and Employment Laws in American Cities* (New York: National Employment Law Project, 2009). Food retail firms are not among the industries with the highest violation rates according to this study, which found that apparel manufacturing, personal services, and household employment all have violation rates that hover in the vicinity of 40 percent.

4. Raj Patel, *Stuffed and Starved: Markets, Power and the Hidden Battle for the World Food System* (London: Portobello Books, 2007).

5. See especially Charles Wollenberg's "Introduction," in John Steinbeck, *The Harvest Gypsies* (Berkeley, CA: Haydey Press, 1998).

6. John Rawls, *The Theory of Justice* (Cambridge, MA: Harvard University Press, 1972).

7. Sidney Mintz, *Sweetness and Power: The Place of Sugar in Modern History* (New York: Penguin Books, 1985).

8. R. Douglas Hurt, *American Agriculture: A Brief History*, rev. ed. (West Lafayette, IN: Purdue University Press, 2002).

9. Carolyn Sachs, *Gendered Fields: Rural Women, Agriculture, and Environment* (Boulder, CO: Westview Press, 1996).

10. Bina Agrawal, *A Field of One's Own: Gender and Land Rights in South Asia* (Cambridge: Cambridge University Press, 1994).

11. Robert Figueroa and Claudia Mills, "Environmental Justice," in *A Companion to Environmental Philosophy*, ed. D. Jamison (Oxford: Basil Blackwell, 2001), 426–438. My rendering here is not totally faithful to Figueroa and Mills, but I hope that they not find it contrary to their views. See also Kristin Shrader-Frechette, *Environmental Justice: Creating Equality, Reclaiming Democracy* (New York: Oxford University Press, 2005); P. Mohai, D. Pellow, and J. T. Roberts, "Environmental Justice," *Annual Review of Environment and Resources* 34 (2009): 405–430.

12. Eric Schlosser, *Fast Food Nation: The Dark Side of the American Meal* (Boston: Houghton-Mifflin, 2001). Several chapters were serialized by *Rolling Stone* in 1999, and the book has been republished several times in succeeding years with additional material.

13. Michael Pollan, *The Omnivore's Dilemma: A Natural History of Four Meals* (New York: Penguin Press, 2004).

14. Julie Guthman, "Commentary on Teaching Food: Why I am Fed Up with Michael Pollan et al.," *Agriculture and Human Values* 24 (2007): 261–264.

15. Paul Rozin, Linda Millman, and Carol Nemeroff, "Operation of the Laws of Sympathetic Magic in Disgust and Other Domains," *Journal of Personality and Social Psychology* 50 (1986): 703–712; Paul Rozin, Claude Fischler, Sumio Imada, Alison Sarubin, and Amy Wrzesniewski, "Attitudes to Food and the Role of Food in Life in the U.S.A., Japan, Flemish Belgium and France: Possible Implications for the Diet-health Debate," *Appetite* 33 (1999): 163–180.

16. Pollan's 2010 review of five books spanning topics in ethical vegetarianism, dietary health, slow food, fairness to food workers, and alternative farming testifies to this phenomenon. Michael Pollan, "The Food Movement, Rising," *The New York Review of Books* 10 (2010): 31–33.

17. EPIC stands for "End Poverty in California." See Greg Mitchell, *The Campaign of the Century: Upton Sinclair's Race for Governor of California and the Birth of Media Politics* (Sausalito, CA: Polipoint Press, 2010).

18. Food and Agricultural Organization, Hunger Portal, http://www.fao.org/hunger/en/, accessed December 15, 2013.

19. Paul B. Thompson, "Food Aid and the Famine Relief Argument (Brief Return)," *The Journal of Agricultural and Environmental Ethics* 23 (2010): 209–227.

20. James A. Harrington, *The Common-Wealth of Oceana* (London: J. Streater for Livewell Chapman, 1656).

21. Miguel Altieri, "Agroecology, Small Farms and Food Sovereignty," *Monthly Review* 61, no. 3 (2009): 102–113; Raj Patel, "What Does Food Sovereignty Look Like?," *Journal of Peasant Studies* 36 (2009): 663–673; William D. Schanbacher, *The Politics of Food: The Global Conflict Between Food Security and Food Sovereignty* (Santa Barbara, CA: ABC-CLIO, 2010).

22. Eric Holt Giménez and Annie Shattuck, "Food Crises, Food Regimes and Food Movements: Rumblings of Reform or Tides of Transformation?," *Journal of Peasant Studies* 38 (2011): 109–144.

23. Michael Pollan, "The Food Movement Rising," *New York Review of Books*, June 10, 2010, http://www.nybooks.com/articles/archives/2010/jun/10/food-movement-rising/, accessed December 15, 2013.

Chapter 3

1. Vegetarianism was apparently advocated by members of Pythagorean school that is believed to be the precursor of Western philosophy in the tradition of Socrates and Plato. A surviving work by Porphyry of Tyre (234–c. 305) provides a systematic argument against the killing and consumption of animals for food. Like today, vegetarianism was advocated both as a healthful practice and as a consequence of ethical duties owed to nonhuman animals.

2. Ivars Avotins, "Training and Frugality in Seneca and Epicurus," *Phoenix* 31 (1977): 214–217. Martha Nussbaum reads Epicurus fragments on food as cautions against adopting false beliefs that could counter the body's "built-in" ability to be satiated. Martha Nussbaum, *The Therapy of Desire: Theory and Practice in Hellenistic Ethics* (Princeton, NJ: Princeton University Press, 1994), 111–112.

3. Orby Shipley, *A Theory about Sin in Relation to Some Facts about Daily Life* (London: Macmillan, 1875), 269.

4. http://en.wikipedia.org/wiki/Gluttony, accessed January 2, 2013.

5. R. F. Yeager, "Aspects of Gluttony in Chaucer and Gower," *Studies in Philology* 81, no. 1 (Winter 1984): 42–55.

6. St. Thomas Aquinas, *The "Summa Theologica" of St. Thomas Aquinas* (London: Burnes, Oates & Washborne, 1913), 2648.

7. Ibid., 2649.

8. The moral importance of dietetic practice was configured in a number of different ways during the medieval era. One view is that each of the mortal sins follows from another in a kind of domino theory of vice. Much like a single puff of marijuana was once thought to lead naturally to harder drugs, then on to addiction and a life of ruin, gluttony might have be thought to be a threshold vice, leading naturally to lust, greed, sloth, wrath, envy, and vainglory. A less mechanistic view sees each of the mortal sins as a form of psychological corruption, a tendency to be overcome by temptation or by passion of any kind. Siegfried Wenzel argues that both Greek and Christian thinkers such as Aquinas saw the moral problem less as one of corruption of the soul than as misdirection of the will. A moral person is focused on the good and the right; conduct motivated by pleasure seeking is morally problematic precisely because it *is not* reflective of the good will. Siegfried Wenzel, "The Seven Deadly Sins: Some Problems of Research," *Speculum* 43 (1968): 1–22.

9. Although Shapin believes that the dietary moderation comes to be seen as the norm, he also notes one peculiar descent from ancient dietary views that associates asceticism with those otherworldly people who attain special wisdom. Thus, while a moderate diet is for ordinary people, it is not surprising when a figure such as Sir Isaac Newton (1642–1726), Robert Boyle (1627–1621), or (closer to our own time) Ludwig Wittgenstein (1889–1951) displays a kind of ethereal contact with the world of ideas through dietary practices of extreme self-denial. In this pattern, ascetic dietary practice has a moral significance, to be sure, but it is a form of spiritual excellence that is bought at the price of bodily health. Stephen Shapin, *Never Pure: Historical Studies of Science as if It Was Produced by People with Bodies, Situated in Time, Space, Culture and Society, and Struggling for Credibility and Authority* (Baltimore: Johns Hopkins University Press, 2010), 237–258.

10. Tristram Stuart, *The Bloodless Revolution: A Cultural History of Vegetarianism From 1600 to Modern Times* (New York: W. W. Norton, 2007).

11. Shapin, *Never Pure*, p. 270.

12. Ibid., 274, 279–281.

13. Ibid., 278.

14. Susan Bordo, *Unbearable Weight: Feminism, Western Culture, and the Body* (Berkeley: University of California Press, 1993).

15. Walter Sinnott-Armstrong, "It's Not My Fault: Global Warming and Individual Moral Obligations," *Advances in the Economics of Environmental Research* 5 (2005): 285–307.

16. Susan Okie, "The Employer as Health Coach," *New England Journal of Medicine* 357.15 (2007): 1465; Michelle M. Mello and Meredith B. Rosenthal, "Wellness Programs and Lifestyle Discrimination—the Legal Limits," *New England Journal of Medicine* 359.2 (2008): 192–199.

17. Tsjalling Swierstra, "Behavior, Environment or Body: Three Discourses on Obesity," in *Genomics, Obesity and the Struggle over Responsibilities*, ed. M. Korthals (Dordrecht, NL: Springer, 2011), 27–38; Michiel Korthals, "Obesity Genomics: Struggle over Responsibilities," in *Genomics, Obesity and the Struggle over Responsibilities*, 77–94.

18. Henk van den Belt, "Contesting the Obesity 'Epidemic': Elements of a Counter Discourse," in *Genomics, Obesity and the Struggle over Responsibilities*, 39–57.

19. Michiel Korthals, "Three Main Areas of Concern, Four Trends in Genomics and Existing Deficiencies in Academic Ethics," in *Genomics, Obesity and the Struggle over Responsibilities*, 59–76; Maartje Schermer, "Genomics, Obesity and Enhancement," in *Genomics, Obesity and the Struggle over Responsibilities*, 131–148. Although I take the point with regard to using the word *epidemic* in this context, I continue to refer to the statistical increase in obesity-related diseases as an epidemic mainly for convenience in exposition.

20. Jennifer Crocker, "Social Stigma and Self-esteem: Situational Construction of Self-worth," *Journal of Experimental Social Psychology* 35.1 (1999): 89–107; Samantha Kwan, "Individual versus Corporate Responsibility: Market Choice, the Food Industry, and the Pervasiveness of Moral Models of Fatness," *Food, Culture and Society: An International Journal of Multidisciplinary Research* 12.4 (2009): 477–495; Kathleen LeBesco, "Neoliberalism, Public Health, and the Moral Perils of Fatness," *Critical Public Health* 21.2 (2011): 153–164.

21. Kristina H. Lewis and Meredith B. Rosenthal, "Individual Responsibility or a Policy Solution—Cap and Trade for the U.S. Diet?," *New England Journal of Medicine* 365, no. 17 (2011): 1561–1563.

22. Thomas Nagel and James Sterba, "Agent-relative Morality," *The Ethics of War and Nuclear Deterrence* (Belmont, CA: Wadsworth, 1985), 15–22; Thomas Nagel, *The View from Nowhere* (New York: Oxford University Press, 1989).

23. Benjamin Caballero, "The Global Epidemic of Obesity: An Overview," *Epidemiologic Reviews* 29.1 (2007): 1–5.

24. J. K. Binkley, J. Eales, and M. Jekanowski, "The Relation between Dietary Change and Rising US Obesity," *The Journal of Obesity* 24 (2000): 1032–1039.

25. Kiyah J. Duffey and Barry M. Popkin, "Shifts in Patterns and Consumption of Beverages between 1965 and 2002," *Obesity* 15 (2007): 2739–2747.

26. Reneé Boynton-Jarrett, Tracy N. Thomas, Karen E. Peterson, Jean Wiecha, Arthur M. Sobol, and Steven L. Gortmaker, "Impact of Television Viewing Patterns on Fruit and Vegetable Consumption Among Adolescents," *Pediatrics* 112, no. 6 (2003): 1321–1326.

27. David A. Kessler, *The End of Overeating: Taking Control of the Insatiable American Appetite* (New York: Rodale Books, 2009).

28. The thesis is examined and rejected by Søren Holm. "Parental Responsibility and Obesity in Children," *Public Health Ethics* 1, no. 1 (2008): 21–29.

29. Ann Vileisis, *Kitchen Literacy: How We Lost the Knowledge of Where Our Food Comes From, and Why We Need to Get It Back* Washington, DC: Island Press, 2008.

30. Meredith Minkler, "Personal Responsibility for Health? A Review of the Arguments and the Evidence at Century's End," *Health Education & Behavior* 26, no. 1 (1999): 121–141.

31. Harriet O. Kunkel, "Nutritional Science at Texas A & M University, 1888–1984," Texas Agricultural Experiment Station Bulletin no. 1490, College Station, June 1985.

32. Joan Dye Gussow, "Improving the American Diet," *Journal of Home Economics* 65, no. 8 (1973): 6–10; *The Feeding Web: Issues in Nutritional Ecology* (Palo Alto, CA: Bull Publishing Company, 1978).

33. Gyorgy Scrinis, "Nutritionism and Functional Foods," in *The Philosophy of Food*, ed. D. Kaplan (Berkeley: University of California Press, 2012), 269–291.

34. Michael Pollan, "Farmer in Chief," *New York Times Magazine* October 12, 2008; also *In Defense of Food: An Eater's Manifesto* (New York: Penguin Books, 2008).

35. Michiel Korthals, "Prevention of Obesity and Personalized Nutrition: Public and Private Health," in *Genomics, Obesity and the Struggle*, 191–205.

36. Albert Lee and Susannah E. Gibbs, "Neurobiology of Food Addiction and Adolescent Obesity Prevention in Low-and Middle-Income Countries," *Journal of Adolescent Health* 52, no. 2 (2013): S39–S42.

37. Hannah Landecker, "Food as Exposure: Nutritional Epigenetics and the New Metabolism," *BioSocieties* 2 (2011): 167–194; James Trosko, "Pre-Natal Epigenetic Influences on Acute and Chronic Diseases Later in Life, Such as Cancer: Global Health Crises Resulting from a Collision of Biological and Cultural Evolution," *Journal of Food Science and Nutrition* 16 (2011): 394–407.

Chapter 4

1. Alisha Coleman-Jensen, Mark Nord, and Anita Singh, *Household Food Security in the United States in 2012*, Economic Research Report No. (ERR-155) 41, September 2013 (Washington, DC: Government Printing Office, 2013).

2. Food and Agriculture Organization, *The State of Food Insecurity in the World 2012* (Rome: Food and Agriculture Organization, 2012), http://www.fao.org/docrep/016/i3027e/i3027e00.htm, accessed August 1, 2014.

3. Peter Singer, "Famine, Affluence and Morality," *Philosophy and Public Affairs* 1 (1972): 229–243.

4. Paul R. Ehrlich, *The Population Bomb* (New York: Ballantine Books, 1968).

5. Garrett Hardin, "The Tragedy of the Commons," *Science* 162 (1968): 1243–1248; "Lifeboat Ethics: The Case against Helping the Poor," *Psychology Today Magazine*, vol. 8, September 1974, 38–43, 123–126; "Carrying Capacity as an Ethical Concept," *Soundings* 58 (1976): 120–137; *The Limits Of Altruism: An Ecologist's View Of Survival* (Bloomington: University of Indiana Press, 1976).

6. Paul R. Ehrlich and Anne H. Ehrlich, "The Population Bomb Revisited," *The Electronic Journal of Sustainable Development* 1, no. 3 (2009), http://mrhartansscienceclass.pbworks.com/w/file/fetch/53321328/The%20Population%20Bomb%20Revisited.pdf, accessed July 5, 2014.

7. Peter Unger, *Living High and Letting Die: Our Illusion of Innocence* (New York: Oxford University Press, 1996); Peter Singer, *Practical Ethics*, 2nd ed. (New York: Cambridge University Press, 1993).

8. Amartya Sen, *Poverty and Famine: An Essay on Entitlement and Deprivation* (New York: Oxford University Press, 1981); Jean Drèze and Amartya Sen, *Hunger and Public Action*, (New York: Oxford University Press, 1989).

9. Food security is absorbed into imperatives for general bodily health in many recent treatments of development ethics. See Thomas Pogge, *World Poverty and Human Rights* (Cambridge, UK: Polity Press, 2007); Martha Nussbaum, *Women and Human Development: The Capabilities Approach* (Cambridge, UK: Cambridge University Press, 2000); Elizabeth Ashford, "The Duties Imposed by the Human Right to Basic Necessities,"

in *Freedom from Poverty as a Human Right: Who Owes What to the Very Poor?*, ed. T. Pogge (Cambridge, UK: Polity Press, 2007), 183–218.

10. This problem has been well known among agricultural economists for over half a century. See Theodore W. Schultz, "Impact and Implications of Foreign Surplus Disposal on Underdeveloped Economies: Value of U.S. Farm Surpluses to Underdeveloped Countries," *American Journal of Agricultural Economics* 42 (1960): 1019–1030.

11. Marcel Mazoyer and Laurence Roudart, *A History of World Agriculture: From the Neolithic Age to The Current Crisis*, translated by J. Membrez (London: Earthscan, 2006).

12. Paul B. Thompson, *The Ethics of Aid and Trade: US Food Policy, Foreign Competition and the Social Contract* (New York: Cambridge University Press, 1992).

13. The statement that farmers are unable to find buyers at any price is a hyperbole that nonetheless communicates a larger economic truth. Markets for farm commodities grow when people make dietary shifts toward consuming meat and other animal products because the animals will consume several pounds of grain for every pound of meat, milk, or eggs. Arguably, shifts toward consumption of simple sugars such as high fructose corn syrup have also broken through the ceiling, resulting in the dietary disaster discussed in Chapter 3. Farmers can also expand markets for farm produce by shifting to "luxury" foods like coffee and tea, or to non-food crops such as cotton or biofuels. But the larger point remains: within a relatively stable market that is localized temporally and geographically, farmers can and do produce more than they can sell.

14. Thompson, *Ethics of Aid and Trade*.

15. The ethical justifiability of Pareto better outcomes has been sharply debated. See Ezra J. Mishan, "The Futility of Pareto-efficient Distributions," *The American Economic Review* (1972): 971–976; Mark Sagoff, "Values and Preferences," *Ethics* (1986): 301–316; Daniel Hausman and Michael McPherson, "Economics, Rationality, and Ethics," in *The Philosophy of Economics: An Anthology*, 2nd ed. (New York: Cambridge University Press, 1994), 252–277.

16. Jeffery Sachs, *The End of Poverty: Economic Possibilities for Our Time* (New York: Penguin Press, 2006).

17. Sach's view is discussed at greater length in my book *The Agrarian Vision*, where I called this the industrial philosophy of

agriculture. The industrial philosophy should not be confused with what is often called "industrial agriculture." Advocates of the industrial philosophy might well endorse smallholder production in developing countries so long as this form of social organization achieves the greatest possible cost efficiency. Similarly, they might endorse organic production methods if consumer demand for organic products warrants it. Paul B. Thompson, *The Agrarian Vision: Sustainability and Environmental Ethics* (Lexington: University Press of Kentucky, 2010).

18. Henry Shue, *Basic Rights: Subsistence, Affluence, and US Foreign Policy* (Princeton, NJ: Princeton University Press, 1980), chapter one.

19. David A. Crocker, *Ethics of Global Development: Agency, Capability and Deliberative Democracy* (Cambridge, UK: Cambridge University Press, 2008).

20. Ibid.

21. See David Crocker and Ingrid Robeyns, "Capability and Agency," in *Amartya Sen*, ed. C. Morris (New York: Cambridge University Press, 2010), 60–90.

22. Thomas Jefferson, *Notes on the State of Virginia*, originally published 1784, http://www.learnnc.org/lp/editions/nchist-newnation/4478, accessed December 16, 2013.

23. Ibid.

Chapter 5

1. Tristram Stuart, *The Bloodless Revolution: A Cultural History of Vegetarianism From 1600 to Modern Times* (New York: W. W. Norton, 2007).

2. Richard Sorabji, *Animal Minds and Human Morals: The Origins of the Western Debate* (Ithaca, NY: Cornell University Press, 1993).

3. This statement of the five freedoms is quoted from the webpage of the Farm Animal Welfare Committee (FAWC), which operates under the Department for Environment, Food and Rural Affairs. The FAWC is a successor to the Farm Animal Welfare Council, which was a "non departmental public body" that was itself a successor to the Brambell Committee operating until 2011. http://www.defra.gov.uk/fawc/about/five-freedoms/, accessed December 19, 2013.

4. Stanley Godlovitch, Roslind Godlovitch, and John Harris, eds., *Animals, Men, and Morals: An Enquiry into the Maltreatment of Non-Humans* (New York: Taplinger Publishing Company, 1972).

5. Tom Regan, *The Case for Animal Rights* (Berkeley: University of California Press, 1983).

6. Bernard E. Rollin, *Farm Animal Welfare: Social, Bioethical and Research Issues* (Ames: Iowa State University Press, 1995).

7. Relatively few of the philosophers who have contested the arguments for ethical vegetarianism have concerned themselves with the actual standards that are or should be applied in contemporary livestock production. One philosopher who has done so is Richard Haynes. Haynes argues that *no* production practice that requires the death of animals *will ever* be morally justifiable. However, given that millions of farm animals will continue to be produced for the foreseeable future, ethics requires that we try to make the conditions in which they live as humane as possible. Richard P. Haynes, "The Myth of Happy Meat," in *The Philosophy of Food*, ed. D. Kaplan (Berkeley: University of California Press, 2012), 161–168.

8. Forty years ago, Nagel speculated that a science would develop which was able to give us insight into the experience of other species. Thomas Nagel, "What Is It Like to Be a Bat?," *The Philosophical Review* 83, no. 4 (1974): 435–450.

9. The argument in this section could be framed in terms of a call for "non-ideal theory" in animal ethics. A recent spate of theorists have argued that classic philosophical theories from Kant and Bentham down through Rawls have relied too heavily on idealizations that distort the moral parameters of the choices that people actually make.

10. I do not mean to imply that this statement is an exhaustive specification of the "freedom from pain" requirement. Husbandry also requires measures to anticipate the potential for injuries, for example, and to take measures that would rectify especially hazardous environments or handling methods. I hope that readers will understand that I am not writing a welfare manual here. The point is simply to show how the Five Freedoms function as measures of relative well-being.

11. The most pro-animal/ethical vegetarian reader of my manuscript wrote, "This paragraph brings up a lot of questions for me." I'm answering some of them in this note on the suspicion that similarly

motivated readers will have the same questions. "1) Are you implying that to engage in answering Q3 we're giving up on thinking animals have rights in favor of welfare?" Answer: Not at all. Trade-offs may be endemic to welfare evaluation, and that's a reason to be careful in moving too quickly from a characterization of welfare or well-being to a rights claim, but I'm not making any claims here about whether or not some kind of rights view could be justified for nonhumans. "2) Mightn't an important part of all the 'freedoms' for human and non-human animals be the ability to have control over these other freedoms?" Answer: Of course, but there are two reasons we don't explore this point here. One is that no creature has total control over their freedoms. All animals, including humans, are subject to some form of social control, sometimes for their own good. I would address the question of *how much* control an individual pig, cow, or chicken should have in those interspecies systems we call farms through the three attributes framework I discuss. The second reason reminds us of the non-ideal nature of question 3. We are just assuming that there will be livestock farms from here on down, and the justification for that assumption was worked out in the preceding section. Some degree of farmer control is implied by that assumption. "3) These problems of 'over-generalizing' might just mean that the five freedoms are inferior to Nussbaum's list of capabilities." Martha Nussbaum has argued that Amartya Sen's capabilities approach (alluded to briefly in Chapter 4) can be extended to nonhuman animals. See Martha C. Nussbaum, *Frontiers of Justice: Disability, Nationality and Species Membership* (Cambridge, MA: Harvard University Press, 2007). Answer: One could say that, in general, capabilities represent an interesting way to bridge aspects of rights-based ethics and utilitarianism. In "Capabilities, Consequentialism and Critical Consciousness," in *Capabilities, Power and Institutions: Toward a More Critical Development Ethics*, ed. Stephen L. Esquith and Fred Gifford (University Park: The Pennsylvania State University Press, 2010), 163–170, I've argued that this was probably Sen's intent. But we are way too deep into the details of a specific ethical theory for an introduction to food ethics.

12. David Fraser, Dan M. Weary, Edward A. Pajor, and Barry N. Milligan, "A Scientific Conception of Animal Welfare that Reflects Ethical Concerns," *Animal Welfare* 6 (1997): 187–205; David Fraser, "Animal Ethics and Animal Welfare Science: Bridging

the Two Cultures," *Applied Animal Behaviour Science* 65, no. 3 (1999): 171–189.

13. See note 7. Of course, there are philosophers and cognitive scientists who deny or sharply qualify this, as alluded to briefly at the beginning of the chapter. If they are right, it is unlikely that raising farm animals presents any ethical difficulty at all. However, farmers themselves have never questioned that animals exhibiting the classic behavioral indication of pain or fear are actually experiencing a mental state of some kind when they do so. We thus leave the more skeptical arguments to the philosophers of mind.

14. Michael C. Appleby, *What Should We Do about Animal Welfare?* (Oxford: Blackwell Science, 1999). As I am developing the framework, it could be applied to any animal that can be said to have a welfare, including humans. Appleby does not include humans, but neither does he specifically exclude them. There may be some animals that do not have welfare (amoebae perhaps?), and some have argued that among food animals crustaceans and mollusks may fall into this category. Whether or not they have feelings, they certainly have veterinary well-being and species-typical behavior. My treatment in this chapter is not meant to exclude such questions from food ethics, though surely they involve subtleties that go well beyond an introductory text. For a more detailed discussion, see Gary E. Varner, *Personhood, Ethics, and Animal Cognition: Situating Animals in Hare's Two Level Utilitarianism* (New York: Oxford University Press, 2012).

15. Appleby, *What Should We Do about Animal Welfare?*.

16. Hence the term *battery cage*. The sometimes quoted suggestion that these cages are the size of an average automobile battery has no basis in fact.

17. Michael C. Appleby, Joy A. Mench, and Barry O. Hughes, *Poultry Behaviour and Welfare* (Cambridge, MA: CABI, 2004).

18. US producers typically use white breeds that are slightly smaller than the brown hens used in many other parts of the world.

19. UEP represents the producers of "shell eggs," which are typically shipped and sold in the shell. There are also producers of liquid and powdered eggs that are used in processed foods as well as by commercial and institutional kitchens (hospitals, restaurants, and hotels). There is no reason to suppose that these producers house hens at more than 48 square inches. Also, in full disclosure, I have served on the United Egg

Producers Animal Welfare Science Advisory Committee since 2005. I was not involved in the recommendation to increase space.

The UEP reform was accomplished by stocking the existing caged production facilities at lower densities, roughly a 25 percent reduction in hens per house. It would have been economic suicide for a single producer to do this on his own initiative because competitors using the 48 square-inch rate could have easily sold their eggs for less that it would have cost the animal Good Samaritan to produce at 68 square inches. By moving as an industry, the UEP created a "level playing field" for producers to make this move voluntarily. Nevertheless, it was not a foregone conclusion that all members would play along. The measure succeeded in part because UEP introduced a "UEP approved" label for producers who complied with new rules, and retailers rapidly adopted a policy of refusing to sell eggs that did not bear the "UEP approved" label.

20. Hens naturally have a sharp beak that can penetrate feathers and break the skin of the pecked bird. In order to limit injuries from pecking, commercial producers trim the sharp point from the beak of the hen when she is still a chick. Beak trimming (also called "de-beaking") is also controversial from a welfare standpoint.

21. But what about broiler chickens raised for their meat? They are typically kept uncaged in flocks of 100,000 or more. Is pecking a problem in broiler production? The answer seems to be no. The reason is that broilers and layers have been bred to be substantially different animals. Exploring the difference also takes us into speculative territory. Broilers grow rapidly and tend to be torpid. One conjecture is that they establish local dominance hierarchies in these large CAFOs: if you pick up a bird and move it to the opposite end of the house it will immediately be pecked to death. A more plausible conjecture is that breeding to increase rapid growth in broilers has also reduced traits tied to aggression and domination, perhaps by increasing "juvenile" characteristics. Correlatively, breeding to increase the rate of egg laying has had the opposite effect. Both genetic correlations have occurred largely by chance. The example shows how a less introductory discussion of farm animal ethics moves into questions about what is permissible or perhaps even obligatory to do with breeding or biotechnology. If we could "solve" the problem of pecking by manipulating a bird's genetics, would we have also "harmed" the bird by altering the genetic basis

for a species-typical behavior? See Paul B. Thompson, "The Opposite of Human Enhancement: Nanotechnology and the Blind Chicken Problem," *NanoEthics* 2 (2008): 305–316.

22. Stanley Cavell, Cora Diamond, John McDowell, and Ian Hacking, *Philosophy and Animal Life* (New York: Columbia University Press, 2008), 131.

23. Regan, *The Case for Animal Rights*, 324–327.

24. Varner, *Personhood, Ethics and Animal Cognition*. Varner goes on to argue against a strict requirement for vegetarianism that is based on his view that personhood is a requirement for the most stringent moral obligations and his assessment of farm animals' cognitive capacities. I am not drawing on this aspect of Varner's theory here.

25. Jamie Pearce, Tony Blakely, Karen Witten, and Phil Bartie, "Neighborhood Deprivation and Access to Fast-Food Retailing: A National Study," *American Journal of Preventive Medicine* 35 (2007): 375–382. I hope that readers will appreciate that I am in no way endorsing the fact of food deserts or recommending that poor people eat at fast-food restaurants. I am only saying that they *do* this, and that, given the constraints under which they live, we can hardly blame them morally for it.

26. The aphorism circulates with the attribution to Aua, but in anthropologist Knud Rasmussen's study of Iglulik Eskimos, the statement is rendered, "The greatest peril of life lies in the fact that human food consists entirely of souls." Rasmussen reports the statement as occurring during a conversation in which Aua is sharing some thoughts on the religious beliefs of his people. However, Rasmussen reports that it was said by Aua's younger brother Ivaluardjuk. Rasmussen's text is almost certainly the definitive source from which popular versions of the aphorism derive. Knud Rasmussen, *Intellectual Culture of the Iglulik Eskimos: Report of the Fifth Thule Expedition, Vol. VII, No. 1* (Copenhagen: Gylendalske Boghandel, Nordisk Forlag, 1929), from a reprint published in New York: AMS Press, 1976, 56.

27. These are the types of question considered by Varner, *Personhood, Ethics and Animal Cognition*.

28. In particular, they do not exhibit elevated cortisol levels. Comparative studies of stall and group housing utilize a number of measures and typically conclude that stalls perform better on some measures of well-being. See A. D. Sorrells, S. D. Eicher, M. J. Harris,

E. A. Pajor, and B. T. Richert, "Periparturient Cortisol, Acute Phase Cytokine, and Acute Phase Protein Profiles of Gilts Housed in Groups or Stalls during Gestation," *Journal of Animal Science* 85 (2007): 1750–1757; Guillermo A. M. Karlen, Paul H. Hemsworth, Harold W. Gonyou, Emma Fabrega, A. David Strom, and Robert J. Smits, "The Welfare of Gestating Sows in Conventional Stalls and Large Groups on Deep Litter," *Applied Animal Behaviour Science* 105 (2007): 87–101.

29. My pro-vegetarian reader writes this comment: "But would humans raised in similar settings show signs of perturbation if they were raised that way? Similarly, would captured wild pigs show signs of perturbation?" Answer: I don't know about humans but I know that wild pigs would, at least, need a significant amount of time to adapt before their stress levels would be remotely comparable. Under normal conditions it would be ethically inhumane to cage wild suidae this way for any length of time. But if wild pigs were experiencing severe food deprivation in nature, they might be quite content to remain in an environment where food was regularly dispensed so that no other animal could get it. And pigs raised on concrete are initially fearful of walking on grass or dirt, so similar logic might imply that *un-caging* them would be inhumane were it not for the fact that we know they adapt quickly. In short, the adaptive ability of pigs and humans means that we *must* ask questions about what makes a good life for these species in a light of the capacities or capabilities that we think they *should* realize as habituated preferences. Chickens may not be so adaptive on an individual basis. Their adaptive ability may reside primarily in their genes. All this is why I take the animal natures part of the framework quite seriously (a point that should trouble anyone inclined to take the view that ethics is about satisfying preferences). As the main text asserts, I am not saying that the absence of physiological stress is a defense of CAFOs. The point here is intended to counter the assumption—which I (perhaps incorrectly) attribute to many nonspecialists—that these pigs are enduring a form of constant suffering tantamount to waterboarding.

30. As noted, beef production may be a partial exception. A significant number of relatively small "cow-calf" operators continue to survive by producing calves that are often eventually sold to "feeders" where they grow to market weight. Some economists believe that many cow-calf producers would not survive but for the fact that they are supported by off-farm income.

31. Almost all farmers perform a mix of manual and managerially complex labor that would be compensated at rates well above that of the average farm household income if were they to enter conventional labor markets. Interestingly, my pro-vegetarian reader also objects to this point, stating that this is not how people normally use the word *profit*. It's true that the ordinary language conventions here are complex. We will ask a kid who ran a lemonade stand for the afternoon (generally at Mom and Dad's expense), "What were your profits?" On the other hand, teachers typically spend some of their own money for supplies and (at the college level) research materials, yet we don't ask them whether *they* made a profit on these expenditures when they collect their paychecks. I'm sticking with what the main text says here. When we get in the realm of ethics, it is just erroneous to think that someone who simply recovers their costs—including their living costs—is making a profit. Economists can be very helpful in giving us criteria for determining when profit-taking has occurred.

32. If the discussion in Chapter 4 did not sufficiently make the point, note again that generalizations on farm income and wealth have many exceptions, especially when farming is considered in its global breadth. Some cattle and sheep ranchers in Latin America are among the wealthiest individuals in their respective societies.

33. Bailey Norwood and Jayson Lusk, *Compassion by the Pound: The Economics of Animal Welfare* (New York: Oxford University Press, 2012).

34. Consider, for example, the growing practice of backyard chickens. Anecdotally, I find that almost all backyard producers lose hens to disease, injury, and predators at a rate that far exceeds the commercial industry's average of 2 percent per year. Although empirical data is scarce, some studies suggest that even among knowledgeable producers, mortality rates will be between five and ten times higher. See P. K. Biswasa, D. Biswasa, S. Ahmeda, A. Rahmanb, and N. C. Debnathc, "A Longitudinal Study of the Incidence of Major Endemic and Epidemic Diseases Affecting Semi-scavenging Chickens Reared under the Participatory Livestock Development Project Areas in Bangladesh," *Avian Pathology* 34 (2005): 303–312; Czech Conroy, Nick Sparks, D. Chandrasekaran, Anshu Sharma, Dinesh Shindey, L. R. Singh, A. Natarajan, and K. Anitha, "Improving Backyard Poultry-Keeping: A Case Study From India," *AG-REN Network Paper* No. 146, UK Department for International Development, 2005;

Patrick J. Kelly, Daniel Chitauro, Christopher Rohde, John Rukwava, Aggrey Majok, Frans Davelaar, and Peter R. Mason, "Diseases and Management of Backyard Chicken Flocks in Chitungwiza, Zimbabwe," *Avian Diseases* 38, no. 3 (July–September 1994): 626–629.

35. Furthermore, all of these producer groups are focused on multiple objectives, lobbying the federal government for favorable trade policy and subsidies not the least of them. So although all of these organizations do *something* about animal welfare, it would be fair to say that redefining the rules of competition in order to accommodate animal interests has not been their highest priority.

36. The 20 percent figure cited was reported to me by several European scientists working on animal welfare issues between 2010 and 2013. More generally, see US Department of Agriculture Foreign Agricultural Service, Global Agricultural Information Network (GAIN) Report E60042, July 4, 2011, http://gain.fas.usda.gov/Recent%20GAIN%20 Publications/Implementation%20of%20Animal%20Welfare%20 Directives%20in%20the%20EU_Brussels%20USEU_EU-27_7-14-2011.pdf, accessed December 22, 2013; Paul Ingenbleek, Victor M. Immink, Hans AM Spoolder, Martien H. Bokma, and Linda J. Keeling, "EU Animal Welfare Policy: Developing a Comprehensive Policy Framework," *Food Policy* 37, no. 6 (2012): 690–699.

37. Heng Wei Cheng and W. M. Muir, "Mechanisms of Aggression and Production in Chickens: Genetic Variations in the Functions of Serotonin, Catecholamine, and Corticosterone," *World's Poultry Science Journal* 63, no. 2 (2007): 233–254; Heng Wei Cheng, "Breeding of Tomorrow's Chickens to Improve Well-being," *Poultry Science* 89, no. 4 (2010): 805–813.

38. The philosophical literature on these extreme technological approaches is far more extensive than the literature on basic changes in husbandry. See Bernard Rollin, *The Frankenstein Syndrome: Ethical and Social Issues in the Genetic Engineering of Animals* (New York: Cambridge University Press, 1995); Peter Sandøe, Birte Lindstrøm Nielsen, Lars Gjøl Christensen, and P. Sorensen, "Staying Good while Playing God: The Ethics of Breeding Farm Animals," *Animal Welfare* 8, no. 4 (1999): 313–328; Bernice Bovenkerk, Frans WA Brom, and Babs J. Van den Bergh. "Brave New Birds: The Use of 'Animal Integrity' in Animal Ethics," *Hastings Center Report* 32, no. 1 (2002): 16–22. Varner, *Personhood, Ethics, and Animal Cognition*, note 145 offers an ethical argument in favor of these approaches, but

he also recognizes that consumer resistance makes pursuing them somewhat unrealistic.

39. But I do not claim that reducing harms does anything much to rectify the larger ethical issue of humanity's instrumental use of other animals, up to and including their death. On this point, see Leonard Lawlor, *This Is Not Sufficient: An Essay on Animality and Human Nature in Derrida* (New York: Columbia University Press, 2007).

40. Yet another issue in dietary ethics arises in connection with the trials and tribulations that vegetarians endure from those with more conventional eating habits. The issue is discussed in Chapter 6 of *The Agrarian Vision*.

Chapter 6

1. Richard J. Hobbs, Eric S. Higgs, and Carol M. Hall, "Defining Novel Ecosystems," In *Novel Ecosystems: Intervening in the New Ecological World Order*, ed. R. J. Hobbs, E. S. Higgs and C. M. Hall (Hoboken, NJ: John Wiley and Sons, 2013), 58.

2. Of course, it also depends on what one means by *pristine*. See Thomas R. Vale, "The Pre-European Landscape of the United States: Pristine or Humanized?," in *Fire, Native Peoples, and the Natural Landscape*, ed. Thomas R. Vale (Washington, DC: The Island Press, 2002), 1–40.

3. Laura Westra, "A Transgenic Dinner: Social and Ethical Issues in Biotechnology and Agriculture," *Journal of Social Philosophy* 24 (1993): 213–232.

4. Rachel Carson, *Silent Spring* (Boston: Houghton Mifflin, 1962).

5. World Commission on Environment and Development (WCED), *Our Common Future* (New York: Oxford University Press, 1987).

6. An *allele* is one among several configurations of the same gene. Natural differences in eye or hair color are due to the presence of several distinct alleles for the genes that determine pigmentation in the human gene pool. See my book *Food Biotechnology in Ethical Perspective* for more detailed discussions of the rationale and philosophical problems associated with biodiversity conservation. Paul B. Thompson, *Food Biotechnology in Ethical Perspective*, 2nd ed. (Dordrecht, NL: Springer, 2007). For an exhaustive discussion, see Donald S. Maier, *What's So Good about Biodiversity? A Call for Better Reasoning about Nature's Value* (Dordrecht, NL: Springer, 2012).

7. Gordon Douglass, "The Meanings of Agricultural Sustainability," in *Agricultural Sustainability in a Changing World Order*, ed. G. K. Douglass (Boulder, CO: Westview Press, 1984), 3–30.

8. The diagram was intended to support Altieri's claim that it would be crucial for agriculture to be sustainable in all three of these domains. Although it is far from clear how to reconcile Douglass's three domains with Altieri's diagram, this way of thinking about sustainability in the context of agriculture has had enduring influence. Altieri may well have been the first author to use this now familiar "three pillars" or "three circles" approach to sustainability. Miguel Altieri, *Agroecology: The Scientific Basis of Sustainable Agriculture* (Boulder, CO: Westview Press, 1987).

9. The analysis in terms of resource sufficiency and functional integrity summarizes an argument that is developed in considerably greater detail in Paul B. Thompson, *The Agrarian Vision: Sustainability and Environmental Ethics* (Lexington: University Press of Kentucky, 2010).

10. WCED, *Our Common Future*, 43.

11. Intergovernmental Panel on Climate Change, *Climate Change 2013: The Physical Basis* (Cambridge, UK: Cambridge University Press, 2013).

12. Simon Dresner, *The Principles of Sustainability*, 2nd ed. (New York: Earthscan, 2008).

13. Thompson, *The Agrarian Vision*.

14. Attributed to Bentham by John Stuart Mill, *Utilitarianism*. See J. M. Robson et al., eds., *Collected Works of John Stuart Mill* (Toronto/London: University of Toronto Press, 1963–1991), vol. 10, p. 257.

15. This is an admittedly unsubtle account of the non-identity problem. Readers with an interest in the ethics of future generations are advised to consult the voluminous literature that has been generated in response to Derek Parfit, *Reasons and Persons* (New York: Oxford University Press, 1984).

16. I am, of course, taking some liberties in my summary treatment of a work that runs to several hundred pages. See G. W. F. Hegel, *Philosophy of History*, translated by J. Sibree (Mineola, NY: Dover, 1956).

17. Marcel Mazoyer and Lawrence Roudart, *A History of World Agriculture from the Neolithic to the Present Age*, translated by James H. Membrez (London: Earthscan, 2006).

18. Aristotle writes that there are three divisions to the *polis*: the very rich (*agathoi*), the very poor (*kakoi*), and the middle (*mesoi*). He applies his principle of moderation to suggest that a *polis* based on this middle class is best. See Aristotle, *Politics*, translated by Benjamin Jowett (New York: Modern Library, 1943), Book Four, Part XI.

19. Victor Davis Hanson, *The Other Greeks: The Family Farm and the Roots of Western Civilization*, 2nd ed. (Berkeley: University of California Press, 1999).

20. Ibid.

21. The claim that peppered moths exemplify Darwinian evolution has been debated. See Fiona Proffitt, "In Defense of Darwin and a Former Icon of Evolution," *Science* 304 (2004): 1894–1895. However, this debate is not material to the question of whether changes in the moth's color are an instance of adaptation within a population, which is the point I am illustrating here.

22. Holmes Rolston, III, *Genes, Genesis and God: Values and their Origins in Natural and Human History* (Cambridge, UK: Cambridge University Press, 1999).

23. Clifford W. Hesseltine, John J. Ellis, and Odette L. Shotwell, "Helminthosporium: Secondary Metabolites, Southern Leaf Blight of Corn, and Biology," *Journal of Agricultural and Food Chemistry* 19 (1971): 707–717.

24. The T-cytoplasm example is not, in my view, enough to prove that American corn production is unsustainable, even in a functional integrity sense. Perhaps a close examination of incentive structures would show that it is quite resilient, even when it is not particularly robust. Farmers are, after all, still growing plenty of maize in Iowa some fifty years later. It is not easy to figure out how we should bound the systems that we associate with functional integrity, and further inquiry into this question must await the work of other authors.

25. I take it that the essays in Lance H. Gunderson and C. S. Holling, eds., *Panarchy: Understanding Transformations in Human and Natural Systems* (Washington, DC: Island Press, 2002) were an important attempt to think through the boundary-setting questions, but I do not find these authors to have developed a coherent approach to the boundaries question in functional integrity. In *The Agrarian Vision* I have argued that the virtues orientation of traditional agrarian philosophies has unappreciated advantages.

26. Ann Vileisis, *Kitchen Literacy: How We Lost the Knowledge of Where Our Food Comes From, and Why We Need to Get It Back* (Washington, DC: Island Press, 2008).

27. Lappé's book is discussed in Chapter 1. Readers who are willing to work through a careful analysis of the environmental and climate impact of various forms of livestock production are advised to consult Simon Fairlie, *Meat: A Benign Extravagance* (White River Junction, VT: Chelsea Green, 2010). For a brief and representative sample of places where one can find variations of this argument in print, consider Jill M. Dieterle, "Unnecessary Suffering," *Environmental Ethics* 30 (2008): 51–67; Jan Dekkers, "Vegetarianism: Sentimental or Ethical?," *Journal of Agricultural and Environmental Ethics* 22 (2009): 573–597; Ramona Cristina Ilea, "Intensive Livestock Farming: Global Trends, Increased Environmental Concerns, and Ethical Solutions," *Journal of Agricultural and Environmental Ethics* 22.2 (2009): 153–167; Anthony McMichael and Ainslie J. Butler, "Environmentally Sustainable and Equitable Meat Consumption in a Climate Change World," in *The Meat Crisis: Developing More Sustainable Production and Consumption*, ed. Joyce D'Silva and John Webster (London: Earthscan, 2010), 173–189; Leanne Bourgeois, "A Discounted Threat: Environmental Impacts of the Livestock Industry," *Earth Common Journal* 2 (2012), no page numbering; Patrick Curry, *Ecological Ethics: An Introduction*, 2nd ed. (Cambridge, UK: Polity Press, 2011); Drew Leder, "Old McDonald's Had a Farm: The Metaphysics of Factory Farming," *Journal of Animal Ethics* 2 (2012): 73–86; Brian G. Henning, "Standing in Livestock's 'Long Shadow': The Ethics of Eating Meat on a Small Planet," *Ethics & the Environment* 16, no. 2 (2011): 63–93; Arianna Ferrari, "Animal Disenhancement for Animal Welfare: The Apparent Philosophical Conundrums and the Real Exploitation of Animals. A Response to Thompson and Palmer," *NanoEthics* 6.1 (2012): 65–76. Stellan Welin ties this argument to the case for synthetic meat: "Introducing the New Meat. Problems and Prospects," *Etikk i praksis* 1.1 (2013), while historian James McWilliams is sharply critical of the localvore ethic, but uses a version of the argument to support a pescatarian ethic on environmental and animal welfare grounds: James E. McWilliams, *Just Food: How Locavores Are Endangering the Future of Food and How We Can Eat Responsibly* (New York: Little Brown, 2009).

28. Henning Steinfeld, Pierre Gerber, Tom Wassenaar, Vincent Castel, Mauricio Rosales, and Cees De Haan, *Livestock's long*

shadow: Environmental Issues and Options (Rome: Food and Agriculture Organization, 2006). Fairlie (*Meat: A Benign Extravagance*, Note 27, this chapter) provides a detailed analysis and exegesis of this point. See also Maurice E. Pitesky, Kimberly R. Stackhouse, and Frank M. Mitloehner, "Clearing the Air: Livestock's Contribution to Climate Change," *Advances in Agronomy* 103 (2009): 1–40. As usual, however, things are not so simple. Other research suggests that the FAO report authors have not adequately taken the carbon sequestration of grasslands into account. See J. F. Soussana, T. Allec, and V. Blanfort, "Mitigating the Greenhouse Gas Balance of Ruminant Production Systems through Carbon Sequestration in Grassland," *Animal* 4 (2010): 334–340.

29. Michael J VandeHaar and Norman St-Pierre, "Major Advances in Nutrition: Relevance to the Sustainability of the Dairy Industry," *Journal of Dairy Science* 89, no. 4 (2006): 1280–1291; Judith L. Capper, "The Environmental Impact of Beef Production in the United States: 1977 Compared with 2007," *Journal of Animal Science* 89, no. 12 (2011): 4249–4261; "Is the Grass Always Greener? Comparing the Environmental Impact of Conventional, Natural and Grass-Fed Beef Production Systems," *Animals* 2 (2012): 127–143.

30. Rich Pirog, Timothy Van Pelt, Kanmar Enshayan, and Ellen Cook, *Food, Fuel, and Freeways: An Iowa Perspective on How Far Food Travels, Fuel Usage, and Greenhouse Gas Emissions* (Ames, IA: The Leopold Center for Sustainable Agriculture, 2001), http://www.leopold. iastate.edu/pubs-and-papers/2001-06-food-fuel-freeways#sthash. ORJYKcKN.dpuf, accessed August 25, 2013. The Department of Energy document is cited by the Iowa State authors as follows: "U.S. Department of Energy. U.S. Agriculture: Potential Vulnerabilities. Menlo Park, CA. 1969, Stanford Research Institute." A personal communication from Rich Pirog, an author of the Iowa State paper, is the basis for my claim that their intention was to stimulate thinking rather than to suggest a valid measure of environmental impact. The idea of food miles was popularized in some of Michael Pollan's early food system journalism. See Helen C. Wagenvoord, "The High Price of Cheap Food," *San Francisco Chronicle Magazine*, May 2, 2004, http:// michaelpollan.com/profiles/the-high-price-of-cheap-food-mealpolitik-over-lunch-with-michael-pollan/, accessed August 25, 2011.

31. Jules N. Pretty, Andy S. Ball, Tim Lang, and James IL Morison, "Farm Costs and Food Miles: An Assessment of the Full Cost of the UK

Weekly Food Basket," *Food Policy* 30, no. 1 (2005): 1–19; Christopher L. Weber and H. Scott Matthews, "Food-Miles and the Relative Climate Impacts of Food Choices in the United States," *Environmental Science and Toxicology* 42 (2008): 3508–3513; Tara Garnett, "Where are the Best Opportunities for Reducing Greenhouse Gas Emissions in the Food System (Including the Food Chain)?," *Food Policy* 36 (2011): S23–S32.

Chapter 7

1. The GM controversy itself is discussed from a social science perspective in *Biotechnology: The Making of a Global Controversy*, edited by M. W. Bauer and G. Gaskell (Cambridge, UK: Cambridge University Press, 2002), and from an agricultural science perspective in Per Pinstrup-Andersen and Ebbe Schiøler, eds., *Seeds of Contention: World Hunger and the Global Controversy Over GM Crops* (Baltimore: Johns Hopkins University Press, 2000). Michael Ruse and David Castle have edited a collection of essays entitled *Genetically Modified Foods: Debating Biotechnology* (Amherst, NY: Prometheus Press, 2002) that collects a useful sample of the pro and con positions discussed above. This anthology includes a number of representative papers on the debate over golden rice, including Ingo Potrykus's "Golden Rice and the Greenpeace Dilemma," which I take to be a model of what insiders *should* do. Richard Sherlock and John Morrey are the editors of a more comprehensive collection entitled *Ethical Issues in Biotechnology* (Lanham, MD: Rowman and Allenheld, 2002) that also includes papers on topics in medical ethics.

2. Noah Zerbe, "Feeding the Famine? American Food Aid and the GMO Debate in Southern Africa," *Food Policy* 29 (2004): 593–608.

3. Dear Reader: If the lesson you took from Chapter 4 is that cheaper food is always a bad thing, please go back and read again. Reducing the cost of food is generally a good thing for people who buy their food. Technology that creates farm-level efficiencies (hence lowering the cost of food) can force some producers out of farming. But many (not all) of the poorest farmers do need to get more production per unit of labor if they are ever to escape poverty.

4. Louise W. Knight, *Jane Addams: Spirit in Action* (New York: W. W. Norton, 2010).

5. I do not think that taking a rights approach precludes one from reaching an equivalent position, but I leave this argument in its utilitarian framing in order to simplify an already complex analysis.

6. Borlaug's views have been aired on at least one television broadcast (Bill Moyers's NOW, October 2002) and published in a number of outlets, including an editorial in the February 6, 2000, *Wall Street Journal* ("We Need Biotech to Feed the World") and a longer 2001 article published as "Ending World Hunger. The Promise of Biotechnology and the Threat of Antiscience Zealotry," in *Plant Physiology* 124: 487–490, and also in the *Transactions* of the Wisconsin Academy of Arts and Sciences, Volume 89 (2001).

7. The Green Revolution is itself a highly contested effort to benefit the poor by improving their technology. The term *Green Revolution* was apparently coined in 1968 by a US Agency for International Development (USAID) administrator named William Gaud (see William S. Gaud, "The Green Revolution: Accomplishments and Apprehensions," *AgBioWorld*, 2011 [1968], http://www.agbioworld. org/biotech-info/topics/borlaug/borlaug-green.html. It is now widely and indiscriminately used to indicate deployment of industrial era agricultural technology in a developing world context. The 1960s-era technologies emphasized dwarf varieties that were more capable of converting the energy from fertilizer into grain. The first generation of Green Revolution crops centered on conventionally bred dwarf varieties of wheat and rice that were widely adopted throughout Asia and Latin America. These new seeds were distributed free of charge through international development agencies funded by national governments (such as USAID) and the Rockefeller Foundation.

However, to be useful these varieties required ready access to fertilizer. Green Revolution farmers were thus induced into a pattern of debt finance for annual fertilizer purchases. Fertilizer distributors also offered pesticides and commercial seeds, leading to an overall transformation of peasant agriculture. Gender played a significant role, as on-farm decision making was typically done by women, but trade and contracts (such as incursion of debts and input purchases) were dominated by men. Succinctly, the Green Revolution crops precipitated a cascade of social and economic transformations throughout peasant agriculture, and introduced a set of technological practices that had already been tied to significant environmental impact in the developed world. These impacts were well understood within the development community by 1980, and long before the advent of biotechnology. See Walter P. Falcon, "The Green Revolution: Generations of Problems," *American Journal of Agricultural Economics* 52, no. 5 (1970): 698–710; Harry M. Cleaver, "The Contradictions of the Green Revolution," *The American Economic Review* 62, no. 1/2 (1972): 177–186; Keith Griffin, *The Political Economy of Agrarian Change, An Essay on the Green Revolution* (London: Macmillan, 1974).

8. Peter Singer, "Famine, Affluence and Morality," *Philosophy and Public Affairs* 1 (19720: 229–243.

9. Ibid.

10. Robert Paarlberg, *Starved for Science: How Biotechnology is Being Kept Out of Africa* (Cambridge, MA: Harvard University Press, 2008). In Chapter 8, I argue that Paarlberg's defense is a bit too passionate. In addition to the papers by Norman Borlaug cited above, the Borlaug hypothesis can be found in articles such as Anthony Trewavas's, "Much Food, Many Problems," *Nature* 17 (1999): 231–232, and Florence Wambugu's "Why Africa Needs Agricultural Biotech," *Nature* 400 (1999): 15–16. Both of these papers are in the Ruse and Castle volume. Philosopher Gregory Pence makes a longer but still very broad case for the Borlaug hypothesis in his book *Designer Genes: Mutant Harvest or Breadbasket of the World?* (Lanham, MD: Rowman & Littlefield, 2002). Each of these works is notable for the paucity of references to or discussion of published books and articles by critics of biotechnology. Gary Comstock has developed a philosophically sophisticated version of the argument implied by the Borlaug hypothesis in his book *Vexing Nature: On the Ethical Case Against Agricultural Biotechnology* (Boston: Kluwer Academic Publishers, 2000), and more concise version of his argument can be found in Gary Comstock, "Ethics and Genetically Modified Foods," http://scope.educ.washington.edu/ gmfood/commentary/show.php?author=Comstock, which is also included in the Ruse and Castle volume.

11. Norman E. Borlaug, "Sixty-two Years of Fighting Hunger: Personal Recollections," *Euphytica* 157, no. 3 (2007): 287–297; Leon F. Hesser, *The Man Who Fed The World: Nobel Peace Prize Laureate Norman Borlaug and His Battle To End World Hunger: An Authorized Biography* (Dallas: Durban House Publishing Company, 2006).

12. Arguments against agricultural biotechnology were discussed at some length in my book *Food Biotechnology in Ethical Perspective*, 2nd ed. (Dordrecht, NL: Springer, 2007), and more succinctly in

Paul B. Thompson and William Hannah, "Food and Agricultural Biotechnology: A Summary and Analysis of Ethical Concerns," *Advances in Biochemical Engineering and Biotechnology* 111 (2008): 229–264.

Empirical support for the contention that ethical arguments reside at the core of even risk-based concerns about GM crops can be found in Paul Sparks, Roger Shepherd, and Lynn Frewer, "Gene Technology, Food Production and Public Opinion: A U.K. Study," *Agriculture & Human Values* 11 (1994): 19–28; Lynn J. Frewer, Roger Shepherd, and Paul Sparks, "Public Concerns in the United Kingdom about General and Specific Aspects of Genetic Engineering: Risk, Benefit and Ethics," *Science, Technology & Human Values* 22 (1997): 98–124. See also John Durant, Martin W. Bauer, and George Gaskell, eds., *Biotechnology in the Public Sphere* (London: The Science Museum, 1998); George Gaskell and Martin W. Bauer, eds., *Biotechnology: The Years of Controversy* (London: The Science Museum, 2001); and Susanna Priest, *A Grain of Truth* (Lanham, MD: Rowman and Littlefield, 2001).

13. Henk Van den Belt, "Debating the Precautionary Principle: 'Guilty until Proven Innocent' or 'Innocent until Proven Guilty'?," *Plant Physiology* 132, no. 3 (2003): 1122–1126; Gary Comstock, "Ethics and Genetically Modified Foods," in *Food Ethics*, ed. F-T Gottwald, H. W. Ingensiep, and M. Meinhardt (New York: Springer, 2010), 49–66.

14. Carl Cranor, *Regulating Toxic Substances: A Philosophy of Science and the Law* (New York: Oxford University Press, 1993).

15. Ann Vileisis, *Kitchen Literacy: How We Lost the Knowledge of Where Our Food Comes From, and Why We Need to Get It Back* (Washington, DC: Island Press, 2008).

16. Linda C. Cummings, "The Political Reality of Artificial Sweeteners," in *Consuming Fears: The Politics of Product Risks*, ed. H. M. Sapolsky (New York: Basic Books, 1986), 116–140.

17. Joseph D. Rosen, "Much Ado about Alar," *Issues in Science and Technology* 7, no. 1 (1990): 85–90.

18. Bruce N. Ames and Lois Swarsky Gold, "Pesticides, Risk and Applesauce," *Science* 244 (1989): 755–757.

19. In truth, regulatory protection for ecosystems on such grounds is quite weak, largely because the methods for documenting benefits and risks of agriculture (as opposed to pesticide use) are not well developed. See Scott M. Swinton, Frank Lupi, G. Philip Robertson, and Stephen

K. Hamilton, "Ecosystem Services and Agriculture: Cultivating Agricultural Ecosystems for Diverse Benefits," *Ecological Economics* 64 (2007): 245–252.

20. Perhaps the best overall resource on the conceptual basis for the precautionary principle is Joel Ticknor and Carolyn Raffensperger, eds., *Protecting Public Health and the Environment: Implementing the Precautionary Principle* (Washington, DC: Island Press, 1999), although this volume does not discuss GM crops. The Ruse and Castle volume includes a paper by Florence Dagicour entitled "Protecting the Environment: from Nucleons to Nucleotides," but other papers in the volume by Indur Goklany and by Henry Miller and Gregory Conko are critical of the precautionary approach as it is generally understood by its advocates. Among the philosophically sophisticated publications advocating its use with respect to biotechnology, see Carl F. Cranor, "How Should Society Approach the Real and Potential Risks Posed by New Technologies?," *Plant Physiology* 133 (2003): 3–9 and Michiel Korthals, "Ethics of Differences in Risk Perception and Views on Food Safety," *Food Protection Trends* 24, no. 7 (2004): 30–35.

21. Willard Cochrane, *The Development of American Agriculture: A Historical Analysis* (Minneapolis: University of Minnesota Press, 1979).

22. Jack Kloppenburg, Jr., *First the Seed: The Political Economy of Plant Biotechnology, 1492–2000* (Cambridge, UK: Cambridge University Press, 1988).

23. Pat Roy Mooney, *The Law of the Seed: Another Development and Plant Genetic Resources* (Uppsala, SE: Dag Hammarskjold Foundation, 1983); Calestus Juma, *The Gene Hunters: Biotechnology and the Scramble for Seeds* (Princeton, NJ: Princeton University Press, 1988).

24. Vandana Shiva, *The Violence of Green Revolution: Third World Agriculture, Ecology and Politics* (London: Zed Books, 1991); *Monocultures of the Mind: Perspectives on Biodiversity and Biotechnology* (London: Palgrave Macmillan, 1993). See Note 7 in this chapter for earlier criticisms of the Green Revolution.

25. Vandana Shiva, *Biopiracy: The Plunder of Nature and Knowledge* (Cambridge, MA: South End Press, 1997). The report from the United Kingdom's Nuffield Council on Bioethics (http://www.nuffieldbioethics.org/) discusses social justice in detail, and the 2003 follow-up report, *The Use of Genetically Modified Crops in Developing*

Countries, is especially relevant. The Nuffield Council studies are support the analysis in the paper in claiming that social justice issues are constraints on the implementation of GM crops rather than arguments against them. More targeted, specific, and recent studies include David Magnus, "Intellectual Property and Agricultural Biotechnology: Bioprospecting or Biopiracy?," in *Who Owns Life?*, ed. D. Magnus and G. McGee (Amherst, NY: Prometheus Books, 2002), 265–276, and Maarten J. Chrispeels, "Biotechnology and the Poor," *Plant Physiology* 124 (2000): 3–6. The Ruse and Castle volume addresses this topic both through the golden rice papers and through one by Robert Tripp, "Twixt Cup and Lip: Biotechnology and Resource Poor Farmers."

26. Prince Charles's radio address is reprinted in Ruse and Castle, and an editorial "My 10 Fears for GM Food" appeared in the June 1, 1999, edition of *The Daily Mail*.

27. Philosophical statements to the effect that biotechnology might be "unnatural" owe a debt to literature in medical ethics. Leon Kass, the chairman of President George W. Bush's advisory group on bioethics wrote a plea against cloning entitled "The Wisdom of Repugnance," *The New Republic*, June 2, 1997, 17–26. Kass's article provides an argument that can be applied broadly, not solely to human beings, and the argument has indeed been adapted specifically to the case of genetically engineered food in Mary Midgley, "Biotechnology and Monstrosity," *The Hastings Center Report* 30, no. 5 (2000): 7–15. A similar but slightly toned-down line of argument can be found in Ruth Chadwick, "Novel, Natural, Nutritious: Towards a Philosophy of Food," *Proceedings of the Aristotelian Society* (2000): 193–208. Perhaps the following article is among the most radical in articulating the view that genetically engineered food is unnatural: Jochen Bockmühl, "A Goethean View of Plants: Unconventional Approaches," in *Intrinsic Value and Integrity of Plants in the Context of Genetic Engineering*, ed. D. Heaf and J. Wirz (Llanystumdwy, UK: International Forum for Genetic Engineering, 2001), 26–31. Other articles in this collection are worth checking out for those who hope to find a basis for thinking biotechnology unnatural.

28. See Gary Comstock, "Is it Unnatural to Genetically Engineer Plants?," *Weed Science* 46 (1998): 647–651, as well as the Comstock contribution to Ruse and Castle. Michael J. Reiss and Roger Straughn

provide a very thorough discussion in *Improving Nature: The Science and Ethics of Genetic Engineering* (Cambridge, UK: Cambridge University Press, 1996). Arguments that biotechnology might be unnatural are also ridiculed and rebutted by Bernard Rollin in *The Frankenstein Syndrome* (Cambridge, UK: Cambridge University Press, 1995). One of the most widely reprinted articles is Mark Sagoff's "Biotechnology and the Natural," *Philosophy and Public Policy Quarterly* 21 (2001): 1–5.

29. Paul B. Thompson, "Why Food Biotechnology Needs an Opt Out," in *Engineering the Farm: Ethical and Social Aspects of Agricultural Biotechnology*, ed. B. Bailey and M. Lappé (Washington, DC: Island Press, 2002), 27–44.

30. Philosophical discussions of autonomy and the rights of consumers can be found in Debra Jackson, "Labeling Products of Biotechnology: Towards Communication and Consent," *Journal of Agricultural and Environmental Ethics* 12 (2000): 319–330; Thompson, "Why Food Biotechnology Needs an Opt Out"; Robert Streiffer and Alan Rubel, "Democratic Principles and Mandatory Labeling of Genetically Engineered Food," *Public Affairs Quarterly* 18 (2004): 205–222; and Paul Weirich, ed., *Labeling Genetically Modified Food: The Philosophical and Legal Debate* (New York: Oxford University Press, 2008). Although I am unaware of philosophers who take a classically utilitarian view of the food choice issue, the utilitarian viewpoint on labeling and choice is frequently taken by people who do not even seem to be aware that they are doing so. See, for example, Donna U. Vogt, *Food Biotechnology in the United States: Science Regulation, and Issues* (Washington, DC: Congressional Research Service, 1999), Order Code RL30198.

31. The "reductionism" theme has been an important element of Vandana Shiva's critique, especially in a relatively early collection that she edited with Ingunn Moser entitled *Biopolitics: A Feminist and Ecological Reader on Biotechnology* (London: Zed Books, 1995). See especially Shiva's concluding essay "Beyond Reductionism." The reductionism theme has been continued especially forcefully in a book by Finn Bowring, *Science, Seeds and Cyborgs: Biotechnology and the Appropriation of Life* (London: Verso Press, 2003).

32. Aretaic objections to food biotechnology are evident in Brewster Kneen's "A Naturalist Looks at Agricultural Biotechnology," in *Engineering the Farm: Ethical and Social Aspects of Agricultural Biotechnology*, ed. B. Bailey and M. Lappé (Washington, DC: Island Press, 2002), 45–60, and such arguments are nicely analyzed by

Ronald Sandler, "An Aretaic Objection to Agricultural Biotechnology," *Journal of Agricultural and Environmental Ethics* 17 (2004): 301–317, which includes further citations to aretaic arguments in the literature. See also his book *Character and Environment: A Virtue-oriented Approach to Environmental Ethics* (New York: Columbia University Press, 2005).

33. I discuss the "feedback loop" that I mention here in the final chapter of *Food Biotechnology in Ethical Perspective*. One can see it in action in an article by Sheldon Krimsky, "Risk Assessment and Regulation of Bioengineered Food Products," *International Journal of Biotechnology* 2 (2002): 31–238. More broadly, the idea that risk and trust are closely correlated is now fairly well established in risk studies. See Douglas Powell and William Leiss, *Mad Cows and Mother's Milk: The Perils of Poor Risk Communication* (Montreal: McGill-Queens University Press, 1997), for its treatment of issues relating to risk perception, political participation and trust. The book has a chapter on GM food that was published with a note indicating that the text was felt to be unduly accepting of the pro-GM point of view by the graduate research assistants who assisted Powell and Leiss.

34. Comstock, "Ethics and Genetically Modified Food," (2010).

Chapter 8

1. Nancy Tuana, "The Speculum of Ignorance: The Women's Health Movement and Epistemologies of Ignorance," *Hypatia* 21, no. 3 (2006): 1–19. Even a cursory summary of the work in feminist epistemologies exceeds the scope of the present work. As a starting point, I direct readers who wish achieve familiarity with this work to the anthology *Feminist Epistemologies*, edited with an introduction by Linda Alcoff and Elizabeth Potter (New York: Routledge, 2013).

2. See, for example, Ayodele Samuel Jegede, "What Led to the Nigerian Boycott of the Polio Vaccination Campaign?," *PLoS Medicine* 4, no. 3 (2007): e73; Wen L. Kilama, "Ethical Perspective on Malaria Research for Africa," *Acta tropica* 95, no. 3 (2005): 276–284; Marcia Angell, "The Ethics of Clinical Research in the Third World," *New England Journal of Medicine* 337 (1997): 847–849.

3. Vickie L. Shavers, Charles F. Lynch, and Leon F. Burmeister, "Knowledge of the Tuskegee Study and its Impact on the Willingness to Participate in Medical Research Studies," *Journal of the National Medical Association* 92, no. 12 (2000): 563–572.

4. Tarla Rae Peterson, *Sharing the Earth: The Rhetoric of Sustainable Development* (Columbia: University of South Carolina Press, 1997).

5. I would be remiss if I overlooked the book of Hugh Lehman, who patiently works through a number of ways in which we might develop and apply standards of rationality in thinking about our food system. Lehman's commitment to the idea of rationality is greater than mine, but he and I arrive at many of the same conclusions about agriculture. See Hugh Lehman, *Rationality and Ethics in Agriculture* (Moscow: University of Idaho Press, 1995).

6. J. S. Petterson, "Perception vs. Reality of Radiological Impact: The Goiânia Model," *Nuclear News* 31, no. 14 (1988): 84–90. The case is also discussed in Jeanne X. Kasperson, Roger E. Kasperson, Nick Pidgeon, and Paul Slovic, "The Social Amplification of Risk: Assessing Fifteen Years of Research and Theory," in *The Social Amplification of Risk*, ed. N. Pidgeon, R. E. Kasperson, and P. Slovic (New York: Cambridge University Press, 2003), 13–46.

7. All these claims can be found to one degree or another in Vandana Shiva and Afsar H. Jafri, "Bursting the GM Bubble: The Failure of GMOs in India," *The Ecologist Asia* 11 (2003): 6–14. A better example of how these claims are embedded in Shiva's anti-globalization argument can be found in virtually any of her books written since 1990, but see especially Vandana Shiva, *Earth Democracy: Justice, Sustainability and Peace* (Cambridge, MA: South End Press, 2005). A more recent newspaper editorial sounding these themes is Vandana Shiva, "The Fine Print of the Food Wars," *The Asian Age*, July 16, 2014, http://www.asianage. com/columnists/fine-print-food-wars-538, accessed July 29, 2014.

8. Britt Bailey and Marc Lappé, "Engineered Crops Struggle in West Texas," *Environmental Commons*, http://environmentalcommons.org/ texasbrew.html, accessed July 29, 2014.

9. Barry J. Barnett, "The US Farm Financial Crisis of the 1980s," *Agricultural History* 74 (2000): 366–380; John D. Ragland and Alan L. Berman, "Farm Crisis and Suicide: Dying on the Vine?," *OMEGA—The Journal of Death and Dying* 22, no. 3 (1990): 173–185.

10. Keith Kloor, "The GMO-Suicide Myth," *Issues in Science and Technology* 30, no. 2 (2014): 65–70.

11. For a textbook treatment of endocrine disruption, see Tim J. Evans, "Reproductive Toxicity and Endocrine Disruption," in *Veterinary Toxicology: Basic and Clinical Principles*, 2nd ed., ed Ramesh

Chandra Gupt (Waltham, MA: Academic Press, 2012), 278–318. For a history, see Sheldon Krimsky, *Hormonal Chaos: The Scientific and Social Origins of the Environmental Endocrine Hypothesis* (Baltimore: Johns Hopkins University Press, 2000).

12. Gilles-Éric Séralini, Emilie Clair, Robin Mesnage, Steeve Gress, Nicolas Defarge, Manuela Malatestab, Didier Hennequinc, and Joël Spiroux de Vendômois, "Long Term Toxicity of a Roundup Herbicide and a Roundup-tolerant Genetically Modified Maize," *Food and Chemical Toxicology* 50 (2012): 4221–4231.

13. Chelsea Snell, Aude Bernheim, Jean-Baptiste Bergé, Marcel Kuntz, Gérard Pascal, Alain Paris, and Agnès E. Ricroch, "Assessment of the Health Impact of GM Plant Diets in Long-term and Multigenerational Animal Feeding Trials: A Literature Review," *Food and Chemical Toxicology* 50 (2012): 1134–1148.

14. A. Wallace Hayes, "Retraction Notice to 'Long Term Toxicity of a Roundup Herbicide and a Roundup-tolerant Genetically Modified Maize'," *Food and Chemical Toxicity* 63 (2014): 244. Séralini and coauthors have actively defended their work against critics and have protested the retraction. Hayes responded to their letter as well as to comments of other critics who had alleged that Monsanto Company had influenced the decision to retract. He also clarified the basis of the retraction: "However, to be very clear, it is the entire paper, with the claim that there is a definitive link between GMO and cancer that is being retracted. Dr. Séralini has been very vocal that he believes his conclusions are correct. In our analysis, his conclusions cannot be claimed from the data presented in his paper." Wallace Hayes, "Reply to Letter to the Editor," *Food and Chemical Toxicology* 65 (2014): 394–395. For his part, Séralini has replied to the reply asserting (correctly, I think) that his paper never asserted a definitive link between GM crops and cancer, in the first place. G. E. Séralini, R. Mesnage, N. Defarge, and J. Spiroux de Vendômois, "Conclusiveness of Toxicity Data and Double Standards," *Food and Chemical Toxicology* 69 (2014): 357–359. Meanwhile, the original paper has been republished: Gilles-Éric Séralini, Emilie Clair, Robin Mesnage, Steeve Gress, Nicolas Defarge, Manuela Malatestab, Didier Hennequinc, and Joël Spiroux de Vendômois, "Republished Study: Long-term Toxicity of a Roundup Herbicide and a Roundup-tolerant Genetically Modified Maize," *Environmental Sciences Europe* 26, no. 1 (2014): 14, and a commentary on the continuing controversy appeared in a July 2014 issue

of the prestigious journal *Nature*. Chris Woolston, "Republished Paper Draws Fire," *Nature* 511, no. 7508 (2014): 129–129.

15. Gilles-Éric Séralini, Joël Spiroux De Vendômois, Dominique Cellier, Charles Sultan, Marcello Buiatti, Lou Gallagher, Michael Antoniou, and Krishna R. Dronamraju, "How Subchronic and Chronic Health Effects Can Be Neglected for GMOs, Pesticides or Chemicals," *International Journal of Biological Sciences* 5 (2009): 438–443.

16. Robin Mesnage, Christian Moesch, Rozenn Le Grand Grand, Guillaume Lauthier, Joël Spiroux de Vendômois, Steeve Gress, and Gilles-Éric Séralini, "Glyphosate Exposure in a Farmer's Family," *Journal of Environmental Protection* 3 (2012): 1001–1003.

17. Robin Mesnage and Gilles-Éric Séralini, "The Need for a Closer Look at Pesticide Toxicity during GMO Assessment," in *Practical Food Safety: Contemporary Issues and Future Directions*, ed. R. Bhat and V. M. Gómez-López (Chichester, UK: John Wiley & Sons, 2014).

18. See Stephan A. Schwartz, "The Great Experiment: Genetically Modified Organisms, Scientific Integrity, and National Wellness," *EXPLORE: The Journal of Science and Healing* 9, no. 1 (January–February 2013): 12–16.

19. Henry I. Miller, *Policy Controversy in Biotechnology: An Insider's View* (San Diego: Academic Press, 1997).

20. In addition to the two Thompsons, others who have weighed in with book-length studies include Gary Comstock, Hugh Lacey, Gregory Pence, and Bernard Rollin, while important articles have been written by Matthias Kaiser, Eric Millstone, Mark Sagoff, and too many others to list here.

21. R. Paul Thompson, *Agro-Technology: A Philosophical Introduction* (Cambridge, UK: Cambridge University Press, 2011).

22. Paul B. Thompson, "Food Biotechnology's Challenge to Cultural Integrity and Individual Consent," *Hastings Center Report* 27, no. 4 (July–August 1997): 34–38.

23. Paul B. Thompson, "Why Food Biotechnology Needs an Opt Out," in *Engineering the Farm: Ethical and Social Aspects of Agricultural Biotechnology*, ed. B. Bailey and M. Lappé (Washington, DC: Island Press, 2002), 27–44.

24. The website for the Agricultural Market Service information on Country of Origin Labeling at USDA can be found here: http://www.ams.

usda.gov/AMSv1.0/COOL, accessed August 1, 2014. Country-of-origin labels are not represented as bearing on safety, but research shows that consumers may interpret them in the context of safety. See Maria L. Loureiro and Wendy J. Umberger, "A Choice Experiment Model for Beef: What US Consumer Responses Tell Us about Relative Preferences for Food Safety, Country-of-origin Labeling and Traceability," *Food Policy* 32 (2007): 496–514.The point here is that one government agency disallows labels because there is no link to scientific evidence for safety or nutrition, while another government agency requires them. I should stress that I see absolutely no problem with this. It is a case where con-textual thinking removes problems with putative inconsistency. Our beliefs about food are replete with this kind of inconsistency, and it would be ridiculous to strive toward eliminating such contradictions just for the sake of a philosophical principle.

25. Robert Paarlberg, *Starved for Science: How Biotechnology is Being Kept Out of Africa* (Cambridge, MA: Harvard University Press, 2009).

26. Frank Uekoetter, "Know Your Soil: Transitions in Farmers' and Scientists' Knowledge in Germany," in *Soils and Societies: Perspectives from Environmental History*, ed. J. R. McNeill and V. Winiwarter (Isle of Harris, UK: White Horse Press, 2006), 322–340.

27. Paul B. Thompson, *The Agrarian Vision: Sustainability and Environmental Ethics* (Lexington: University Press of Kentucky, 2010). See especially Chapter 9.

28. Julie Guthman, *Agrarian Dreams: The Paradox of Organic Farming in California* (Berkeley: University Press of California, 2004).

29. Paul B. Thompson, *The Agrarian Vision*.

30. Pamela C. Ronald and Raoul W. Adamchak, *Tomorrow's Table: Organic Farming, Genetics, and the Future of Food* (New York: Oxford University Press, 2008).

BIBLIOGRAPHY

Agrawal, Bina. *A Field of One's Own: Gender and Land Rights in South Asia*. Cambridge, UK: Cambridge University Press, 1994.

Alcoff, Linda, and Elizabeth Potter, eds. *Feminist Epistemologies*. New York: Routledge, 2013.

Altieri, Miguel. *Agroecology: The Scientific Basis of Sustainable Agriculture*. Boulder, CO: Westview Press, 1987.

Altieri, Miguel. "Agroecology, Small Farms and Food Sovereignty." *Monthly Review* 61, No. 3 (2009): 102–113.

Ames, Bruce N., and Lois Swarsky Gold. "Pesticides, Risk and Applesauce." *Science* 244 (1989): 755–757.

Angell, Marcia. "The Ethics of Clinical Research in the Third World. *New England Journal of Medicine* 337 (1997): 847–849.

Appleby, Michael C. *What Should We Do about Animal Welfare?* Oxford, UK: Blackwell Science, 1999.

Appleby, Michael C., Joy A. Mench, and Barry O. Hughes. *Poultry Behaviour and Welfare*. Cambridge, MA: CABI, 2004.

Aristotle. *Aristotle's Politics*. Translated by Benjamin Jowett. New York: Modern Library, 1943.

Ashford, Elizabeth. "The Duties Imposed by the Human Right to Basic Necessities." In *Freedom from Poverty as a Human Right: Who Owes What to the Very Poor?* Edited by T. Pogge, 183–218. Cambridge, UK: Polity Press, 2007.

Avotins, Ivars. "Training and Frugality in Seneca and Epicurus." *Phoenix* 31 (1977): 214–217.

Bailey, Britt, and Marc Lappé. "Engineered Crops Struggle in West Texas." *Environmental Commons*. http://environmentalcommons. org/texasbrew.html. Accessed July 29, 2014.

Barnett, Barry J. "The US Farm Financial Crisis of the 1980s." *Agricultural History* 74 (2000): 366–380.

Bauer, Martin W., John Durant, and George Gaskell, eds. *Biotechnology in the Public Sphere: A European Sourcebook*. London: NMSI Trading Ltd, 1998.

Bauer, M. W., and G. Gaskell, eds. *Biotechnology: The Making of a Global Controversy*. Cambridge, UK: Cambridge University Press, 2002.

Beekman, Volkert. "You Are What You Eat: Meat, Novel Protein Foods, and Consumptive Freedom." *Journal of Agricultural and Environmental Ethics* 12, no. 2 (2000): 185–196.

Belasco, Warren. *Meals to Come: The History of the Future of Food*. Berkeley: University of California Press, 2006.

Belasco, Warren. *Appetite for Change: How the Counter Culture Took On the Food Industry*. 2nd updated ed. Ithaca, NY: Cornell University Press, 2007.

Bernhardt, Annette, Ruth Milkman, Nik Theodore, Douglas Heckathorn, Mirabai Auer, James DeFilippis, Ana Luz González, Victor Narro, Jason Perelshteyn, Diana Polson, and Michael Spiller. *Broken Laws, Unprotected Workers: Violations of Labor and Employment Laws in American Cities*. New York: National Employment Law Project, 2009.

Binkley, J. K., J. Eales, and M. Jekanowski. "The Relation between Dietary Change and Rising US Obesity." *The Journal of Obesity* 24 (2000): 1032–1039.

Biswasa, P. K., D. Biswasa, S. Ahmeda, A. Rahmanb, and N. C. Debnathc. "A Longitudinal Study of the Incidence of Major Endemic and Epidemic Diseases Affecting Semi-scavenging Chickens Reared under the Participatory Livestock Development Project Areas in Bangladesh." *Avian Pathology* 34 (2005): 303–312.

Bockmühl, Jochen. "A Goethean View of Plants: Unconventional Approaches." In *Intrinsic Value and Integrity of Plants in the Context of Genetic Engineering*. Edited by D. Heaf and J. Wirz. Llanystumdwy, 26–31. Llanstumdwy, UK: International Forum for Genetic Engineering, 2001.

Bordo, Susan. *Unbearable Weight: Feminism, Western Culture, and the Body*. Berkeley: University of California Press, 1993.

Borgmann, Albert. *Real American Ethics: Taking Responsibility for Our Country*. Chicago: University of Chicago Press, 2006.

Borlaug, Norman E. "Ending World Hunger. The Promise of Biotechnology and the Threat of Antiscience Zealotry." *Plant Physiology* 124 (2000): 487–490.

Borlaug, Norman E. "We Need Biotech to Feed the World." *Wall Street Journal*, December 6, 2000, A22.

Borlaug, Norman E. "Ending World Hunger. The Promise of Biotechnology and the Threat of Antiscience Zealotry." *Transactions* 89 (2001): 1–13.

Borlaug, Norman E. "Sixty-two Years of Fighting Hunger: Personal Recollections." *Euphytica* 157, no. 3 (2007): 287–297.

Bourgeois, Leanne. "A Discounted Threat: Environmental Impacts of the Livestock Industry." *Earth Common Journal* 2 (2012), no page numbering.

Bovenkerk, Bernice, Frans WA Brom, and Babs J. Van den Bergh. "Brave New Birds: The Use of 'Animal Integrity' in Animal Ethics." *Hastings Center Report* 32, no. 1 (2002): 16–22.

Boynton-Jarrett, Reneé, Tracy N. Thomas, Karen E. Peterson, Jean Wiecha, Arthur M. Sobol, and Steven L. Gortmaker. "Impact of Television Viewing Patterns on Fruit and Vegetable Consumption Among Adolescents." *Pediatrics* 112, no. 6 (2003): 1321–1326.

Brülde, Bengdt, and J. Sandberg. *Hur bör vi handla? Filosofiska tankar om rättvisemärkt, ve-getariskt och ekologiskt.* Stockholm: Thales, 2012.

Butler, Judith, and Joan Wallach Scott. *Feminists Theorize the Political.* New York: Routledge, 1992.

Caballero, Benjamin. "The Global Epidemic of Obesity: An Overview." *Epidemiologic reviews* 29, no. 1 (2007): 1–5.

Capper, Judith L. "The Environmental Impact of Beef Production in the United States: 1977 Compared with 2007." *Journal of Animal Science* 89, no. 12 (2011): 4249–4261.

Capper, Judith L. "Is the Grass Always Greener? Comparing the Environmental Impact of Conventional, Natural and Grass-Fed Beef Production Systems." *Animals* 2 (2012): 127–143.

Carson, Rachel. *Silent Spring.* Boston: Houghton Mifflin, 1962.

Cartwright, Michael S., Francis O. Walker, Jill N. Blocker, Mark R. Schulz, Thomas A. Arcury, Joseph G. Grzywacz, Dana Mora, Haiying Chen, Antonio J. Marín, and Sara A. Quandt. "The Prevalence of Carpal Tunnel Syndrome in Latino Poultry Processing Workers and Other Latino Manual Workers." *Journal of Occupational and Environmental Medicine* 54, no. 2 (2012): 198–201.

Cavell, Stanley, Cora Diamond, John McDowell, and Ian Hacking. *Philosophy and Animal Life.* New York: Columbia University Press, 2008.

Chadwick, Ruth. "Novel, Natural, Nutritious: Towards a Philosophy of Food." *Proceedings of the Aristotelian Society* (2000): 193–208.

Charles, the Prince of Wales. "Reith Lecture 2000." In *Genetically Modified Foods: Debating Biotechnology*. Edited by Michael Ruse and David Castle, 11–15. Amherst, NY: Prometheus Press, 2002.

Charles, the Prince of Wales. "My 10 Fears for GM Food." *The Daily Mail*, June 1, 1999.

Cheng, Heng Wei, and W. M. Muir. "Mechanisms of Aggression and Production in Chickens: Genetic Variations in the Functions of Serotonin, Catecholamine, and Corticosterone." *World's Poultry Science Journal* 63, no. 2 (2007): 233–254.

Cheng, Heng Wei. "Breeding of Tomorrow's Chickens to Improve Well-being." *Poultry Science* 89, no. 4 (2010): 805–813.

Chrispeels, Maarten J. "Biotechnology and the Poor." *Plant Physiology* 124 (2000): 3–6.

Cleaver, Harry M. "The Contradictions of the Green Revolution." *The American Economic Review* 62, no. 1/2 (1972): 177–186.

Cochrane, Willard. *The Development of American Agriculture: A Historical Analysis*. Minneapolis: University of Minnesota Press, 1979.

Coleman-Jensen, Alisha, Mark Nord, and Anita Singh. *Household Food Security in the United States in 2012*. Economic Research Report No. (ERR-155) 41 September 2013. Washington, DC: Government Printing Office, 2013.

Comstock, Gary. "Is it Unnatural to Genetically Engineer Plants?" *Weed Science* 46 (1998): 647–651.

Comstock, Gary. *Vexing Nature: On the Ethical Case Against Agricultural Biotechnology*. Boston: Kluwer Academic Publishers, 2000.

Comstock, Gary. "Ethics and Genetically Modified Foods." In *Genetically Modified Foods: Debating Biotechnology*. Edited by Michael Ruse and David Castle, 88–108. Amherst, NY: Prometheus Press, 2002.

Comstock, Gary. "Ethics and Genetically Modified Foods." In *Food Ethics*. Edited by F-T Gottwald, H. W. Ingensiep, and M. Meinhardt, 49–66. New York: Springer, 2010.

Comstock, Gary. "Ethics and Genetically Modified Foods." http://scope.educ.washington.edu/gmfood/commentary/show.php?author=Comstock.

Conroy, Czech, Nick Sparks, D. Chandrasekaran, Anshu Sharma, Dinesh Shindey, L. R. Singh, A. Natarajan, K. Anitha. "Improving

Backyard Poultry-Keeping: A Case Study from India." *AG-REN Network Paper* No. 146. UK Department for International Development: The Agricultural Research and Extension Network, 2005.

Conway, Gordon. "Open Letter to Greenpeace." In *Genetically Modified Foods: Debating Biotechnology*. Edited by Michael Ruse and David Castle, 63–64. Amherst, NY: Prometheus Press, 2002.

Cranor, Carl F. *Regulating Toxic Substances: A Philosophy of Science and the Law*. New York: Oxford University Press, 1993.

Cranor, Carl F. "How Should Society Approach the Real and Potential Risks Posed by New Technologies?" *Plant Physiology* 133 (2003): 3–9.

Crocker, David A. *Ethics of Global Development: Agency, Capability and Deliberative Democracy*. Cambridge, UK: Cambridge University Press, 2008.

Crocker, David and Ingrid Robeyns. "Capability and Agency." In *Amartya Sen*. Edited by C. Morris, 60–90. New York: Cambridge University Press, 2010.

Crocker, Jennifer. "Social Stigma and Self-esteem: Situational Construction of Self-worth." *Journal of Experimental Social Psychology* 35, no. 1 (1999): 89–107.

Cummings, Linda C. "The Political Reality of Artificial Sweeteners." In *Consuming Fears: The Politics of Product Risks*. Edited by H. M. Sapolsky, 116–140. New York: Basic Books, 1986.

Curry, Patrick. *Ecological Ethics: An Introduction*. 2nd ed. Cambridge, UK: Polity Press, 2011.

Dagicour, Florence. "Protecting the Environment: from Nucleons to Nucleotides." In *Genetically Modified Foods: Debating Biotechnology*. Edited by Michael Ruse and David Castle, 251–264. Amherst, NY: Prometheus Press, 2002.

Dekkers, Jan. "Vegetarianism: Sentimental or Ethical?" *Journal of Agricultural and Environmental Ethics* 22 (2009): 573–597.

Dewey, John. "The Reflex Arc Concept in Psychology." *Psychological Review* 3, no. 4 (1896): 357–370.

Dewey, John. *Human Nature and Conduct: An Introduction to Social Psychology*. New York: Henry Holt and Company, 1922.

Dewey, John. *Logic: The Theory of Inquiry*. New York: Henry Holt and Company, 1938.

Dieterle, Jill M. "Unnecessary Suffering." *Environmental Ethics* 30, no. 1 (2008): 51–67.

Douglass, Gordon. "The Meanings of Agricultural Sustainability." In *Agricultural Sustainability in a Changing World Order*. Edited by G. K. Douglass, 3–30. Boulder, CO: Westview Press, 1984.

Dresner, Simon. *The Principles of Sustainability* 2nd Ed. New York: Earthscan Press, 2008.

Drèze, Jean, and Amartya Sen. *Hunger and Public Action*. New York: Oxford University Press, 1989.

Duffey, Kiyah J., and Barry M. Popkin. "Shifts in Patterns and Consumption of Beverages between 1965 and 2002." *Obesity* 15 (2007): 2739–2747.

Durant, John, Martin W. Bauer, and George Gaskell, eds. *Biotechnology in the Public Sphere*. London: The Science Museum, 1998.

Ehrlich, Paul R. *The Population Bomb*. New York: Ballantine Books, 1968.

Ehrlich, Paul R., and Anne H. Ehrlich. "The Population Bomb Revisited." *The Electronic Journal of Sustainable Development* 1, no. 3 (2009), http://mrhartanssscienceclass.pbworks.com/w/file/fetch/53321328/The%20Population%20Bomb%20Revisited.pdf. Accessed July 5, 2014.

Eichenwald, Kurt, et al. "Biotechnology Food: From the Label to a Debacle." In *Genetically Modified Foods: Debating Biotechnology*. Edited by Michael Ruse and David Castle, 31–40. Amherst, NY: Prometheus Press, 2002.

Esterbrook, Barry. *Tomatoland: How Modern Industrial Agriculture Destroyed Our Most Alluring Fruit*. Kansas City, MO: Andrews McMeel Publishing, 2012.

Evans, Tim J. "Reproductive Toxicity and Endocrine Disruption." In *Veterinary Toxicology: Basic and Clinical Principles*. 2nd ed. Edited by Ramesh Chandra Gupt, 278–318. Waltham, MA: Academic Press, 2012.

Fairlie, Simon. *Meat: A Benign Extravagance*. White River Junction, VT: Chelsea Green Publishing, 2010.

Falcon, Walter P. "The Green Revolution: Generations of Problems." *American Journal of Agricultural Economics* 52, no. 5 (1970): 698–710.

Ferrari, Arianna. "Animal Disenhancement for Animal Welfare: The Apparent Philosophical Conundrums and the Real Exploitation of Animals. A Response to Thompson and Palmer." *NanoEthics* 6, no. 1 (2012): 65–76.

Figueroa, Robert, and Claudia Mills. "Environmental Justice." In *A Companion to Environmental Philosophy*. Edited by D. Jamison, 426–438. Oxford, UK: Basil Blackwell, 2001.

Fink, Deborah. *Cutting into the Meatpacking Line: Workers and Change in the Rural Midwest.* Chapel Hill: University of North Carolina Press, 1998.

Foucault, Michel. *The History of Sexuality.* New York: Vintage Books, 1990.

Foucault, Michel. *Abnormal: Lectures at the Collège de France, 1974–1975.* London: Verso, 2003.

Fraser, David. "Animal Ethics and Animal Welfare Science: Bridging the Two Cultures." *Applied Animal Behaviour Science* 65, no. 3 (1999): 171–189.

Fraser, David, Dan M. Weary, Edward A. Pajor, and Barry N. Milligan. "A Scientific Conception of Animal Welfare that Reflects Ethical Concerns." *Animal Welfare* 6 (1997): 187–205.

Frewer, Lynn J., Roger Shepherd, and Paul Sparks. "Public Concerns in the United Kingdom about General and Specific Aspects of Genetic Engineering: Risk, Benefit and Ethics." *Science, Technology & Human Values* 22 (1997): 98–124.

Garnett, Tara. "Where Are the Best Opportunities for reducing Greenhouse Gas Emissions in the Food System (Including the Food Chain)?" *Food Policy* 36 (2011): S23–S32.

Gaskell, George, and Martin W. Bauer, eds. *Biotechnology: The Years of Controversy* London: The Science Museum, 2001.

Gaud, William S. "The Green Revolution: Accomplishments and Apprehensions." *AgBioWorld.* 2011 [1968]. http://www.agbio-world.org/biotech-info/topics/borlaug/borlaug-green.html.

Giddens, Anthony. *Modernity and Self-identity: Self and Society in the Late Modern Age.* Palo Alto, CA: Stanford University Press, 1991.

Giménez, Eric Holt, and Annie Shattuck. "Food Crises, Food Regimes and Food Movements: Rumblings of Reform or Tides of Transformation?" *Journal of Peasant Studies* 38 (2011): 109–144.

Glacken, Clarence J. *Traces on the Rhodian Shore: Nature and Culture in Western Thought from Ancient Times to the End of the Eighteenth Century.* Berkeley: University of California Press, 1973.

Godlovitch, Stanley, Roslind Godlovitch, and John Harris, eds. *Animals, Men, and Morals: An Enquiry into the Maltreatment of Non-Humans.* New York: Taplinger Publishing Company, 1972.

Goklany, Indur. "Applying the Precautionary Principle to Genetically Modiefied Crops." In *Genetically Modified Foods: Debating*

Biotechnology. Edited by Michael Ruse and David Castle, 265–291. Amherst, NY: Prometheus Press, 2002.

Greenpeace. "Genetically Engineered 'Golden Rice' Is Fool's Gold." In *Genetically Modified Foods: Debating Biotechnology.* Edited by Michael Ruse and David Castle, 52–54. Amherst, NY: Prometheus Press, 2002.

Griffin, Keith. *The Political Economy of Agrarian Change, An Essay on the Green Revolution.* London: Macmillan, 1974.

Guerinot, Mary Lou. "The Green Revolution Strikes Gold." In *Genetically Modified Foods: Debating Biotechnology.* Edited by Michael Ruse and David Castle, 41–44. Amherst, NY: Prometheus Press, 2002.

Gunderson, Lance H., and C. S. Holling, eds. *Panarchy: Understanding Transformations in Human and Natural Systems.* Washington, DC: Island Press, 2002.

Gussow, Joan Dye. "Improving the American Diet," *Journal of Home Economics* 65, no. 8 (1973): 6–10.

Gussow, Joan Dye. *The Feeding Web: Issues in Nutritional Ecology.* Palo Alto, CA: Bull Publishing Co., 1978.

Guthman, Julie. *Agrarian Dreams: The Paradox of Organic Farming in California.* Berkeley: University Press of California, 2004.

Guthman, Julie. "Commentary on Teaching Food: Why I Am Fed Up with Michael Pollan et al." *Agriculture and Human Values* 24 (2007): 261–264.

Habermas, Jürgen. *Moral Consciousness and Communicative Action.* Cambridge, MA: MIT Press, 1999.

Hanson, Victor Davis. *The Other Greeks: The Family Farm and the Roots of Western Civilization.* 2nd ed. Berkeley: University of California Press, 1999.

Hardin, Garrett. "The Tragedy of the Commons." *Science* 162 (1968): 1243–1248.

Hardin, Garrett. Lifeboat Ethics: The Case against Helping the Poor." *Psychology Today Magazine* 8 (September, 1974): 38–43, 123–126.

Hardin, Garrett. "Carrying Capacity as an Ethical Concept." *Soundings* 58 (1976): 120–137.

Hardin, Garrett. *The Limits Of Altruism: An Ecologist's View Of Survival.* Bloomington: University of Indiana Press, 1976.

Harrington, James A. *The Common-Wealth of Oceana.* London: J. Streater for Livewell Chapman, 1656.

Hausman, Daniel, and Michael McPherson. "Economics, Rationality, and Ethics." In *The Philosophy of Economics: An Anthology.* 2nd ed. Edited by Daniel Hausman, 252–277. New York: Cambridge University Press, 1994.

Hayes, A. Wallace. "Retraction Notice to 'Long Term Toxicity of a Roundup Herbicide and a Roundup-tolerant Genetically Modified Maize.'" *Food and Chemical Toxicity* 63 (2014): 244.

Hayes, Wallace. "Reply to Letter to the Editor." *Food and Chemical Toxicology* 65 (2014): 394–395.

Haynes, Richard P. "The Myth of Happy Meat." In *The Philosophy of Food.* Edited by D. Kaplan, 161–168. Berkeley, CA: University of California Press, 2012.

Hegel, G. W. F. *Philosophy of History.* Translated by J. Sibree. Mineola, NY: Dover, 1956.

Henning, Brian G. "Standing in Livestock's 'Long Shadow': The Ethics of Eating Meat on a Small Planet." *Ethics & the Environment* 16, no. 2 (2011): 63–93.

Hesseltine, Clifford W., John J. Ellis, and Odette L. Shotwell. "Helminthosporium: Secondary Metabolites, Southern Leaf Blight of Corn, and Biology." *Journal of Agricultural and Food Chemistry* 19 (1971): 707–717.

Hesser, Leon F. *The Man Who Fed The World: Nobel Peace Prize Laureate Norman Borlaug and His Battle To End World Hunger: An Authorized Biography.* Dallas: Durban House Publishing Company, 2006.

Hobbs, Richard J., Eric S. Higgs, and Carol M. Hall. "Defining Novel Ecosystems." In *Novel Ecosystems: Intervening in the New Ecological World Order.* Edited by R. J. Hobbs, E. S. Higgs, and C. M. Hall, 58. Hoboken, NJ: John Wiley and Sons, 2013.

Holm, Søren. "Parental Responsibility and Obesity in Children." *Public Health Ethics* 1, no. 1 (2008): 21–29.

Hurt, R. Douglas. *American Agriculture: A Brief History.* Rev. ed. West Lafayette, IN: Purdue University Press, 2002.

Ilea, Ramona Cristina. "Intensive Livestock Farming: Global Trends, Increased Environmental Concerns, and Ethical Solutions." *Journal of Agricultural and Environmental Ethics* 22, no 2 (2009): 153–167.

Ingenbleek, Paul, Victor M. Immink, Hans AM Spoolder, Martien H. Bokma, and Linda J. Keeling. "EU Animal Welfare

Policy: Developing a Comprehensive Policy Framework." *Food Policy* 37, no. 6 (2012): 690–699.

Intergovernmental Panel on Climate Change. *Climate Change 2013: The Physical Basis*. Cambridge, UK: Cambridge University Press, 2013.

Jackson, Debra. "Labeling Products of Biotechnology: Towards Communication and Consent." *Journal of Agricultural and Environmental Ethics* 12 (2000): 319–330.

Jefferson, Thomas. *Notes on the State of Virginia*. Originally published 1784. http://www.learnnc.org/lp/editions/nchist-newnation/4478. Accessed December 16, 2013.

Jegede, Ayodele Samuel. "What Led to the Nigerian Boycott of the Polio Vaccination Campaign?" *PLoS Medicine* 4, no. 3 (2007): e73.

Juma, Calestus. *The Gene Hunters: Biotechnology and the Scramble for Seeds*. Princeton, NJ: Princeton University Press, 1988.

Karlen, Guillermo A.M., Paul H. Hemsworth, Harold W. Gonyou, Emma Fabrega, A. David Strom, and Robert J. Smits. "The Welfare of Gestating Sows in Conventional Stalls and Large Groups on Deep Litter." *Applied Animal Behaviour Science* 105 (2007): 87–101.

Kasperson, Jeanne X., Roger E. Kasperson, Nick Pidgeon, and Paul Slovic. "The Social Amplification of Risk: Assessing Fifteen Years of Research and Theory." In *The Social Amplification of Risk*. Edited by N. Pidgeon, R. E. Kasperson, and P. Slovic, 13–46. New York: Cambridge University Press, 2003.

Kass, Leon. "The Wisdom of Repugnance." *The New Republic*, June 2, 1997, 17–26.

Kelly, Patrick J., Daniel Chitauro, Christopher Rohde, John Rukwava, Aggrey Majok, Frans Davelaar, and Peter R. Mason. "Diseases and Management of Backyard Chicken Flocks in Chitungwiza, Zimbabwe." *Avian Diseases* 38, no. 3 (1994): 626–629.

Kessler, David A. *The End of Overeating: Taking Control of the Insatiable American Appetite*. New York: Rodale Books, 2009.

Kilama, Wen L. "Ethical Perspective on Malaria Research for Africa." *Acta tropica* 95, no. 3 (2005): 276–284.

Kloor, Keith. "The GMO-Suicide Myth." *Issues in Science and Technology* 30, no. 2 (2014): 65–70.

Kloppenburg, Jack, Jr. *First the Seed: The Political Economy of Plant Biotechnology, 1492–2000*. Cambridge, UK: Cambridge University Press, 1988.

Kneen, Brewster. "A Naturalist Looks at Agricultural Biotechnology." In *Engineering the Farm: Ethical and Social Aspects of Agricultural Biotechnology.* Edited by B. Bailey and M. Lappé, 45–60. Washington, DC: Island Press, 2002.

Knight, Louise W. *Jane Addams: Spirit in Action.* New York: W.W. Norton, 2010.

Kolb, David A. *Experiential Learning: Experience as the Source of Learning and Development.* Vol. 1. Englewood Cliffs, NJ: Prentice-Hall, 1984.

Kolb, David A., Richard E. Boyatzis, and Charalampos Mainemelis. "Experiential Learning Theory: Previous Research and New Directions." In *Perspectives on Thinking, Learning, and Cognitive Styles.* Edited by Robert J. Sternberg and Li-Fang Zhang, 227–247. Mahwah, NJ: Lawrence Erlbaum Associates, 2001.

Korsmeyer, Caroline. *Making Sense of Taste: Food and Philosophy.* Ithaca, NY: Cornell University Press, 1999.

Korthals, Michiel. "Ethics of Differences in Risk Perception and Views on Food Safety." *Food Protection Trends* 24, no. 7 (2004): 30–35.

Korthals, Michiel. "Obesity Genomics: Struggle over Responsibilities." In *Genomics, Obesity and the Struggle over Responsibilities.* Edited by M. Korthals, 77–94. Dordrecht, NL: Springer, 2011.

Korthals, Michiel. "Prevention of Obesity and Personalized Nutrition: Public and Private Health." In *Genomics, Obesity and the Struggle over Responsibilities.* Edited by M. Korthals, 191–205. Dordrecht, NL: Springer, 2011.

Korthals, Michiel. "Three Main Areas of Concern, Four Trends in Genomics, and Existing Deficiencies in Academic Ethics." In *Genomics, Obesity and the Struggle over Responsibilities.* Edited by M. Korthals, 59–76. Dordrecht, NL: Springer, 2011.

Krimsky, Sheldon. *Hormonal Chaos: The Scientific and Social Origins of the Environmental Endocrine Hypothesis.* Baltimore: Johns Hopkins University Press, 2000.

Krimsky, Sheldon. "Risk Assessment and Regulation of Bioengineered Food Products." *International Journal of Biotechnology* 2 (2002): 31–238.

Kunkel, Harriet O. "Nutritional Science at Texas A&M University, 1888–1984." *Texas Agricultural Experiment Station Bulletin* No. 1490, College Station, TX. June 1985.

Kwan, Samantha. "Individual versus Corporate Responsibility: Market Choice, the Food Industry, and the Pervasiveness of Moral Models

Of Fatness." *Food, Culture and Society: An International Journal of Multidisciplinary Research* 12, no. 4 (2009): 477–495.

Landecker, Hannah. "Food as Exposure: Nutritional Epigenetics and the New Metabolism." *BioSocieties* 2 (2011): 167–194.

Landlahr, Victor H. *You Are What You Eat.* New York: National Nutrition Society, 1942.

Lappé, Frances Moore. *Diet for a Small Planet.* New York: Ballantine Books, 1975.

Lawlor, Leonard. *This Is Not Sufficient: An Essay on Animality and Human Nature in Derrida.* New York: Columbia University Press, 2007.

LeBesco, Kathleen. "Neoliberalism, Public Health, and the Moral Perils of Fatness." *Critical Public Health* 21, no. 2 (2011): 153–164.

Leder, Drew. "Old McDonald's Had a Farm: The Metaphysics of Factory Farming." *Journal of Animal Ethics* 2 (2012): 73–86.

Lee, Albert, and Susannah E. Gibbs. "Neurobiology of Food Addiction and Adolescent Obesity Prevention in Low- and Middle-Income Countries." *Journal of Adolescent Health* 52, no. 2 (2013): S39–S42.

Lehman, Hugh. *Rationality and Ethics in Agriculture.* Moscow: University of Idaho Press, 1995.

Lewis, Kristina H., and Meredith B. Rosenthal. "Individual Responsibility or a Policy Solution—Cap and Trade for the U.S. Diet?" *New England Journal of Medicine* 365, no. 17 (2011): 1561–1563.

Loureiro, Maria L., and Wendy J. Umberger. "A Choice Experiment Model for Beef: What US Consumer Responses Tell Us about Relative Preferences for Food Safety, Country-of-Origin Labeling and Traceability." *Food Policy* 32 (2007): 496–514.

Magnus, David. "Intellectual Property and Agricultural Biotechnology: Bioprospecting or Biopiracy?" In *Who Owns Life?* Edited by D. Magnus and G. McGee, 265–276. Amherst, NY: Prometheus Books, 2002.

Maier, Donald S. *What's So Good about Biodiversity? A Call for Better Reasoning about Nature's Value.* Dordrecht, NL: Springer, 2012.

Mazoyer, Marcel, and Lawrence Roudart. *A History of World Agriculture from the Neolithic to the Present Age.* Translated by James H. Membrez. London: Earthscan, 2006.

McMichael, Anthony, and Ainslie J. Butler. "Environmentally Sustainable and Equitable Meat Consumption in a Climate Change World." In *The Meat Crisis: Developing More Sustainable Production and Consumption.* Edited by Joyce D'Silva and John Webster, 173–189. London: Earthscan, 2010.

McWilliams, James E. *Just Food: How Locavores Are Endangering the Future of Food and How We Can Eat Responsibly.* New York: Little Brown, 2009.

Mello, Michelle M., and Meredith B. Rosenthal. "Wellness Programs and Lifestyle Discrimination—The Legal Limits." *New England Journal of Medicine* 359, no. 2 (2008): 192–199.

Mesnage, Robin, Christian Moesch, Rozenn Le Grand Grand, Guillaume Lauthier, Joël Spiroux de Vendômois, Steeve Gress, and Gilles-Éric Séralini. "Glyphosate Exposure in a Farmer's Family." *Journal of Environmental Protection* 3 (2012): 1001–1003.

Mesnage, Robin, and Gilles-Éric Séralini. *The Need for a Closer Look at Pesticide Toxicity during GMO Assessment, in Practical Food Safety: Contemporary Issues and Future Directions.* Edited by R. Bhat and V. M. Gómez-López. Chichester, UK: John Wiley & Sons, 2014.

Midgley, Mary. "Biotechnology and Monstrosity: Why We Should Pay Attention to the 'Yuk Factor.'" *Hastings Center Report* 30, no. 5 (2000): 7–15.

Mill, John Stuart. *On Liberty.* New York: Library of Liberal Arts, 1956 [1859].

Miller, Henry I. *Policy Controversy in Biotechnology: An Insider's View.* San Diego: Academic Press, 1997.

Miller, Henry, and Gregory Conko. "Precaution without Principle." In *Genetically Modified Foods: Debating Biotechnology.* Edited by Michael Ruse and David Castle, 292–298. Amherst, NY: Prometheus Press, 2002.

Minkler, Meredith. "Personal Responsibility for Health? A Review of the Arguments and the Evidence at Century's End." *Health Education & Behavior* 26, no. 1 (1999): 121–141.

Mintz, Sidney. *Sweetness and Power: The Place of Sugar in Modern History.* New York: Penguin Books, 1985.

Mishan, Ezra J. "The Futility of Pareto-Efficient Distributions." *The American Economic Review* (1972): 971–976.

Mitchell, Greg. *The Campaign of the Century: Upton Sinclair's Race for Governor of California and the Birth of Media Politics.* Sausalito, CA: Polipoint Press, 2010.

Mohai, P., D. Pellow, and J. T. Roberts. "Environmental Justice." *Annual Review of Environment and Resources* 34 (2009): 405–430.

Mooney, Pat Roy. *The Law of the Seed: Another Development and Plant Genetic Resources*. Uppsala, SE: Dag Hammarskjold Foundation, 1983.

Nagel, Thomas. "What Is It Like to Be a Bat?" *The Philosophical Review* 83, no. 4 (1974): 435–450.

Nagel, Thomas. "Poverty and Food: Why Charity Is Not Enough." In *Food Policy: The Responsibility of the United States in Life and Death Choices*. Edited by P. Brown and H. Shue, 54–62. Boston: The Free Press, 1977.

Nagel, Thomas. *The View from Nowhere*. New York: Oxford University Press, 1989.

Nagel, Thomas, and James Sterba. "Agent-Relative Morality." In *The Ethics of War and Nuclear Deterrence*. Belmont, CA: Wadsworth, 1985.

Norgaard, Kari Marie, Ron Reed, and Carolina Van Horn. "A Continuing Legacy: Institutional Racism, Hunger and Nutritional Justice in the Klamath." In *Cultivating Food Justice: Race, Class and Sustainability*. Edited by A. H. Alkon and J. Agyeman, 32–46. Cambridge, MA: MIT Press, 2011.

Norwood, Bailey, and Jayson Lusk. *Compassion by the Pound: The Economics of Animal Welfare*. New York: Oxford University Press, 2012.

Nuffield Council on Bioethics (UK). *Genetically Modified Crops: The Ethical and Social Issues*. London: Nuffield Council 1999.

Nuffield Council on Bioethics (UK). *The Use of Genetically Modified Crops in Developing Countries: A Follow-up Discussion Paper*. London: Nuffield Council, 2003.

Nussbaum, Martha. *The Therapy of Desire: Theory and Practice in Hellenistic Ethics*. Princeton, NJ: Princeton University Press, 1994.

Nussbaum, Martha. *Women and Human Development: The Capabilities Approach*. Cambridge, UK: Cambridge University Press, 2000.

Nussbaum, Martha C. *Frontiers of Justice: Disability, Nationality and Species Membership*. Cambridge, MA: Harvard University Press, 2007.

Okie, Susan. "The Employer as Health Coach." *New England Journal of Medicine* 357, no. 15 (2007): 1465.

Oliver, Kelly. *Animal Lessons: How They Teach Us to Be Human*. New York: Columbia University Press, 2009.

O'Neill, Onora. *Faces of Hunger*. London: G. Allen & Unwin, 1986.

Paarlberg, Robert. *Starved for Science: How Biotechnology Is Being Kept Out of Africa.* Cambridge, MA: Harvard University Press, 2009.

Parfit, Derek. *Reasons and Persons.* New York: Oxford University Press, 1984.

Patel, Raj. *Stuffed and Starved: Markets, Power and the Hidden Battle for the World Food System.* London: Portobello Books, 2008.

Patel, Raj. "What Does Food Sovereignty Look Like?" *Journal of Peasant Studies* 36 (2009): 663–673.

Pearce, Jamie, Tony Blakely, Karen Witten, and Phil Bartie. "Neighborhood Deprivation and Access to Fast-food Retailing: A National Study." *American Journal of Preventive Medicine* 35 (2007): 375–382.

Pence, Gregory. *Designer Genes: Mutant Harvest or Breadbasket of the World?* Lanham, MD: Rowman & Littlefield, 2002.

Peterson, Tarla Rae. *Sharing the Earth: The Rhetoric of Sustainable Development.* Columbia: University of South Carolina Press, 1997.

Petterson, J. S. "Perception vs. Reality of Radiological Impact: The Goiânia Model." *Nuclear News* 31, no 14 (1988): 84–90.

Pinstrup-Andersen, Per, and Ebbe Schiøler, eds. *Seeds of Contention: World Hunger and the Global Controversy Over GM Crops.* Baltimore: Johns Hopkins University Press, 2000.

Pirog, Rich, Timothy Van Pelt, Kanmar Enshayan, and Ellen Cook. *Food, Fuel, and Freeways: An Iowa Perspective on How Far Food Travels, Fuel Usage, and Greenhouse Gas Emissions.* Ames, IA: The Leopold Center for Sustainable Agriculture, 2001. http://www.leopold.iastate.edu/pubs-and-papers/2001-06-food-fuel-freeways#sthash.ORJYKcKN.dpuf. Accessed August 25, 2013.

Pitesky, Maurice E., Kimberly R. Stackhouse, and Frank M. Mitloehner. "Clearing the Air: Livestock's Contribution to Climate Change." *Advances in Agronomy* 103 (2009): 1–40.

Place, Sara E., and Frank M. Mitloehner. "Contemporary Environmental Issues: A Review of the Dairy Industry's Role in Climate Change and Air Quality and the Potential of Mitigation through Improved Production Efficiency." *Journal of Dairy Science* 93, no. 8 (2010): 3407–3416.

Place, Sara E., and Frank M. Mitloehner. "Beef Production in Balance: Considerations for Life Cycle Analyses." *Meat Science* 92, no. 3 (2012): 179–181.

Pogge, Thomas. *World Poverty and Human Rights*. Cambridge, UK: Polity Press, 2007.

Pollan, Michael. *The Omnivore's Dilemma: A Natural History of Four Meals*. New York: Penguin Press, 2004.

Pollan, Michael. "Farmer in Chief." *New York Times Magazine*, October 12, 2008.

Pollan, Michael. *In Defense of Food: An Eater's Manifesto*. New York: Penguin Books, 2008.

Pollan, Michael. "The Food Movement, Rising." *The New York Review of Books* 10 (2010): 31–33.

Potrykus, Ingo. "Golden Rice and the Greenpeace Dilemma." In *Genetically Modified Foods: Debating Biotechnology*. Edited by Michael Ruse and David Castle, 55–57. Amherst, NY: Prometheus Press, 2002.

Powell, Douglas, and William Leiss. *Mad Cows and Mother's Milk: The Perils of Poor Risk Communication*. Montreal: McGill-Queens University Press, 1997.

Pretty, Jules N., Andy S. Ball, Tim Lang, and James IL Morison. "Farm Costs and Food miles: An Assessment of the Full Cost of the UK Weekly Food Basket." *Food Policy* 30, no. 1 (2005): 1–19.

Priest, Susanna. *A Grain of Truth*. Lanham, MD: Rowman and Littlefield, 2001.

Proffitt, Fiona. "In Defense of Darwin and a Former Icon of Evolution." *Science* 304 (2004): 1894–1895.

Ragland, John D., and Alan L. Berman. "Farm Crisis and Suicide: Dying on the Vine?" *OMEGA—The Journal of Death and Dying* 22, no. 3 (1990): 173–185.

Rasmussen, Knud. *Intellectual Culture of the Iglulik Eskimos: Report of the Fifth Thule Expedition, Vol. VII. No. 1*. Copenhagen: Gylendalske Boghandel, Nordisk Forlag, 1929. From a reprint published in New York: AMS Press, 1976.

Rawls, John. *The Theory of Justice*. Cambridge, MA: Harvard University Press, 1972.

Rawls, John. *Political Liberalism*. New York: Columbia University Press, 1993.

Regan, Tom. *The Case for Animal Rights*. Berkeley: University of California Press, 1983.

Reiss, Michael J., and Roger Straughn. *Improving Nature: The Science and Ethics of Genetic Engineering*. Cambridge, UK: Cambridge University Press, 1996.

Robson, J. M. et al., eds. *Collected Works of John Stuart Mill.* Toronto/London: University of Toronto Press, 1963–1991.

Rollin, Bernard E. *Farm Animal Welfare: Social, Bioethical and Research Issues.* Ames: Iowa State University Press, 1995.

Rollin, Bernard. *The Frankenstein Syndrome: Ethical and Social Issues in the Genetic Engineering of Animals.* New York: Cambridge University Press, 1995.

Rolston, Holmes, III. *Genes, Genesis and God: Values and their Origins in Natural and Human History.* Cambridge, UK: Cambridge University Press, 1999.

Ronald, Pamela C., and Raoul W. Adamchak. *Tomorrow's Table: Organic Farming, Genetics, and the Future of Food.* New York: Oxford University Press, 2008.

Rosen, Joseph D. "Much Ado about Alar." *Issues in Science and Technology* 7, no. 1 (1990): 85–90.

Rozin, Paul, Linda Millman, and Carol Nemeroff. "Operation of the Laws of Sympathetic Magic in Disgust and Other Domains." *Journal of Personality and Social Psychology* 50 (1986): 703–712.

Rozin, Paul, Claude Fischler, Sumio Imada, Alison Sarubin, and Amy Wrzesniewski. "Attitudes to Food and the Role of Food in Life in the U.S.A., Japan, Flemish Belgium and France: Possible Implications for the Diet-Health Debate." *Appetite* 33 (1999): 163–180.

Ruse, Michael, and David Castle, eds. *Genetically Modified Foods: Debating Biotechnology.* Amherst, NY: Prometheus Press, 2002.

Sachs, Carolyn. *Gendered Fields: Rural Women, Agriculture, and Environment.* Boulder, CO: Westview Press, 1996.

Sachs, Jeffrey. *The End of Poverty: Economic Possibilities for Our Time.* New York: Penguin Press, 2006.

Sagoff, Mark. "Values and Preferences." *Ethics* (1986): 301–316.

Sagoff, Mark. "Biotechnology and the Natural," *Philosophy and Public Policy Quarterly* 21 (2001): 1–5.

Sandler, Ronald. "An Aretaic Objection to Agricultural Biotechnology." *Journal of Agricultural and Environmental Ethics* 17 (2004): 301–317.

Sandler, Ronald. *Character and Environment: A Virtue-oriented Approach to Environmental Ethics.* New York: Columbia University Press, 2005.

Sandøe, Peter, Birte Lindstrøm Nielsen, Lars Gjøl Christensen, and P. Sorensen. "Staying Good while Playing God: The Ethics of Breeding Farm Animals." *Animal Welfare* 8, no. 4 (1999): 313–328.

Schanbacher, William D. *The Politics of Food: The Global Conflict Between Food Security and Food Sovereignty.* Santa Barbara, CA: ABC-CLIO, 2010.

Schermer, Maartje. "Genomics, Obesity and Enhancement." In *Genomics, Obesity and the Struggle over Responsibilities.* Edited by M. Korthals, 131–148. Dordrecht, NL: Springer, 2010.

Schlosser, Eric. *Fast Food Nation: The Dark Side of the American Meal.* Boston: Houghton-Mifflin, 2001.

Schultz, Theodore W. "Impact and Implications of Foreign Surplus Disposal on Underdeveloped Economies: Value of U.S. Farm Surpluses to Underdeveloped Countries." *American Journal of Agricultural Economics* 42 (1960): 1019–1030.

Schwartz, Stephan A. "The Great Experiment: Genetically Modified Organisms, Scientific Integrity, and National Wellness." *EXPLORE: The Journal of Science and Healing* 9, no. 1 (January–February 2013): 12–16.

Scrinis, Gyorgy. "Nutritionism and Functional Foods." In *The Philosophy of Food.* Edited by D. Kaplan, 269–291. Berkeley: University of California Press, 2012.

Sen, Amartya. *Poverty and Famine: An Essay on Entitlement and Deprivation.* New York: Oxford University Press, 1981.

Séralini, Gilles-Éric, Joël Spiroux De Vendômois, Dominique Cellier, Charles Sultan, Marcello Buiatti, Lou Gallagher, Michael Antoniou, and Krishna R. Dronamraju. "How Subchronic and Chronic Health Effects Can Be Neglected for GMOs, Pesticides or Chemicals." *International Journal of Biological Sciences* 5 (2009): 438–443.

Séralini, Gilles-Éric, Emilie Clair, Robin Mesnage, Steeve Gress, Nicolas Defarge, Manuela Malatestab, Didier Hennequinc, and Joël Spiroux de Vendômois. "Long Term Toxicity of a Roundup Herbicide and a Roundup-tolerant Genetically Modified Maize." *Food and Chemical Toxicology* 50 (2012): 4221–4231.

Séralini, Gilles-Éric, Emilie Clair, Robin Mesnage, Steeve Gress, Nicolas Defarge, Manuela Malatestab, Didier Hennequinc, and Joël Spiroux de Vendômois. "Republished Study: Long-Term Toxicity of a Roundup Herbicide and a Roundup-tolerant Genetically Modified Maize." *Environmental Sciences Europe* 26, no. 1 (2014): 14.

Séralini, Gilles-Éric, Robin Mesnage, Nicolas Defarge, and Joël Spiroux de Vendômois. "Conclusiveness of Toxicity Data and Double Standards." *Food and Chemical Toxicology* 69 (2014): 357–359.

Shapin, Stephen. *Never Pure: Historical Studies of Science as if It Was Produced by People with Bodies, Situated in Time, Space, Culture and Society, and Struggling for Credibility and Authority.* Baltimore: Johns Hopkins University Press, 2010.

Sharma, Kirti, and Manish Paradakar. "The Melamine Adulteration Scandal." *Food Security* 2, no. 1 (2010): 97–107.

Shavers, Vickie L., Charles F. Lynch, and Leon F. Burmeister. "Knowledge of the Tuskegee Study and Its Impact on the Willingness to Participate in Medical Research Studies." *Journal of the National Medical Association* 92, no. 12 (2000): 563–572.

Sherlock, Richard, and John Morrey, eds. *Ethical Issues in Biotechnology.* Lanham, MA: Rowman and Allenheld, 2002.

Shipley, Orby. *A Theory about Sin in Relation to Some Facts about Daily Life.* London: Macmillan, 1875.

Shiva, Vandana. *The Violence of Green Revolution: Third World Agriculture, Ecology and Politics.* London: Zed Books, 1991.

Shiva, Vandana. *Monocultures of the Mind: Perspectives on Biodiversity and Biotechnology.* London: Palgrave Macmillan, 1993.

Shiva, Vandana. *Biopiracy: The Plunder of Nature and Knowledge.* Cambridge, MA: South End Press, 1997.

Shiva, Vandana. "Golden Rice Hoax: When Public Relations Replace Science." In *Genetically Modified Foods: Debating Biotechnology.* Edited by Michael Ruse and David Castle, 58–62. Amherst, NY: Prometheus Press, 2002.

Shiva, Vandana. "Beyond Reductionism." In *Science, Seeds and Cyborgs: Biotechnology and the Appropriation of Life.* Edited by Finn Bowring. London: Verso Press, 2003.

Shiva, Vandana. *Earth Democracy: Justice, Sustainability and Peace.* Cambridge, MA: South End Press, 2005.

Shiva, Vandana. "The Fine Print of the Food Wars." *The Asian Age*, July 16, 2014, http://www.asianage.com/columnists/fine-print-food-wars-538, accessed July 29, 2014.

Shiva, Vandana, and Ingunn Moser. *Biopolitics: A Feminist and Ecological Reader on Biotechnology.* London: Zed Books, 1995.

Shiva, Vandana, and Afsar H. Jafri. "Bursting the GM Bubble: The Failure of GMOs in India." *The Ecologist Asia* 11 (2003): 6–14.

Shrader-Frechette, Kristin. *Environmental Justice: Creating Equality, Reclaiming Democracy.* New York: Oxford University Press, 2005.

Shue, Henry. *Basic Rights: Subsistence, Affluence, and US Foreign Policy.* Princeton, NJ: Princeton University Press, 1980.

Singer, Peter. "Famine, Affluence and Morality." *Philosophy and Public Affairs* 1 (1972): 229–243.

Singer, Peter. *Practical Ethics*. 2nd ed. New York: Cambridge University Press, 1993.

Singer, Peter. *The Life You Can Save: How to Do Your Part to End World Poverty*. New York: Random House, 2010.

Sinnott-Armstrong, Walter. "It's Not My Fault: Global Warming and Individual Moral Obligations." *Advances in the Economics of Environmental Research* 5 (2005): 285–307.

Snell, Chelsea, Aude Bernheim, Jean-Baptiste Bergé, Marcel Kuntz, Gérard Pascal, Alain Paris, and Agnès E. Ricroch. "Assessment of the Health Impact of GM Plant Diets in Long-Term and Multigenerational Animal Feeding Trials: A Literature Review." *Food and Chemical Toxicology* 50 (2012): 1134–1148.

Sorabji, Richard. *Animal Minds and Human Morals: The Origins of the Western Debate*. Ithaca, NY: Cornell University Press, 1993.

Sorrells, A. D., S. D. Eicher, M. J. Harris, E. A. Pajor, and B. T. Richert. "Periparturient Cortisol, Acute Phase Cytokine, and Acute Phase Protein Profiles of Gilts Housed in Groups or Stalls during Gestation." *Journal of Animal Science* 85 (2007): 1750–1757.

Sparks, Paul, Roger Shepherd, and Lynn Frewer. "Gene Technology, Food Production and Public Opinion: A U.K. Study." *Agriculture & Human Values* 11 (1994): 19–28.

Steinfeld, Henning, Pierre Gerber, Tom Wassenaar, Vincent Castel, Mauricio Rosales, and Cees de Haan. *Livestock's Long Shadow: Environmental Issues and Options*. Rome: Food and Agriculture Organization, 2006.

Streiffer, Robert, and Alan Rubel. "Democratic Principles and Mandatory Labeling of Genetically Engineered Food." *Public Affairs Quarterly* 18 (2004): 205–222.

Stuart, Tristram. *The Bloodless Revolution: A Cultural History of Vegetarianism From 1600 to Modern Times*. New York: W. W. Norton, 2007.

Soussana, J. F., T. Allec, and V. Blanfort, "Mitigating the Greenhouse Gas Balance of Ruminant Production Systems through Carbon Sequestration in Grassland," *Animal* 4 (2010): 334–340.

Swierstra, Tsjalling. "Behavior, Environment or Body: Three Discourses on Obesity." In *Genomics, Obesity and the Struggle over Responsibilities*. Edited by M. Korthals, 27–38. Dordrecht, NL: Springer, 2011.

Swinton, Scott M., Frank Lupi, G. Philip Robertson, and Stephen K. Hamilton. "Ecosystem Services and Agriculture: Cultivating

Agricultural Ecosystems for Diverse Benefits." *Ecological Economics* 64 (2007): 245–252.

Thomas Aquinas. *The "Summa Theologica" of St. Thomas Aquinas.* London: Burnes, Oates & Washborne, 1913.

Thompson, Paul B. *The Ethics of Aid and Trade: US Food Policy, Foreign Competition and the Social Contract.* New York: Cambridge University Press, 1992.

Thompson, Paul B. "Food Biotechnology's Challenge to Cultural Integrity and Individual Consent." *Hastings Center Report* 27, no. 4 (July–August 1997): 34–38

Thompson, Paul B. "Why Food Biotechnology Needs an Opt Out." In *Engineering the Farm: Ethical and Social Aspects of Agricultural Biotechnology.* Edited by B. Bailey and M. Lappé, 27–44. Washington, DC: Island Press, 2002.

Thompson, Paul B. *Food Biotechnology in Ethical Perspective.* 2nd ed. Dordrecht, NL: Springer, 2007.

Thompson, Paul B. "The Opposite of Human Enhancement: Nanotechnology and the Blind Chicken Problem." *NanoEthics* 2 (2008): 305–316.

Thompson, Paul B. *The Agrarian Vision: Sustainability and Environmental Ethics.* Lexington: University Press of Kentucky, 2010.

Thompson, Paul B. "Capabilities, Consequentialism and Critical Consciousness." In *Capabilities, Power and Institutions: Toward a More Critical Development Ethics.* Edited by Stephen L. Esquith and Fred Gifford, 163–170. University Park: Pennsylvania State University Press, 2010.

Thompson, Paul B. "Food Aid and the Famine Relief Argument (Brief Return)." *The Journal of Agricultural and Environmental Ethics* 23 (2010): 209–227.

Thompson, Paul B., and William Hannah. "Food and Agricultural Biotechnology: A Summary and Analysis of Ethical Concerns." *Advances in Biochemical Engineering and Biotechnology* 111 (2008): 229–264.

Thompson, R. Paul. *Agro-Technology: A Philosophical Introduction.* Cambridge, UK: Cambridge University Press, 2011.

Ticknor, Joel, and Carolyn Raffensperger, eds. *Protecting Public Health and the Environment: Implementing the Precautionary Principle.* Washington, DC: Island Press, 1999.

Trewavas, Anthony. "Much Food, Many Problems." *Nature* 17 (1999): 231–232.

Trewavas, Anthony. "Much Food, Many Problems." In *Genetically Modified Foods: Debating Biotechnology*. Edited by Michael Ruse and David Castle, 335–342. Amherst, NY: Prometheus Press, 2002.

Tripp, Robert. "Twixt Cup and Lip: Biotechnology and Resource Poor Farmers." In *Genetically Modified Foods: Debating Biotechnology*. Edited by Michael Ruse and David Castle, 301–303. Amherst, NY: Prometheus Press, 2002.

Trosko, James. "Pre-Natal Epigenetic Influences on Acute and Chronic Diseases Later in Life, Such as Cancer: Global Health Crises Resulting from a Collision of Biological and Cultural Evolution." *Journal of Food Science and Nutrition* 16 (2011): 394–407.

Tuana, Nancy. "The Speculum of Ignorance: The Women's Health Movement and Epistemologies of Ignorance." *Hypatia* 21, no. 3 (2006): 1–19.

Uekoetter, Frank. "Know Your Soil: Transitions in Farmers' and Scientists' Knowledge in Germany." In *Soils and Societies: Perspectives from Environmental History*. Edited by J. R. McNeill and V. Winiwarter, 322–340. Isle of Harris, UK: White Horse Press, 2006.

Unger, Peter. *Living High and Letting Die: Our Illusion of Innocence*. New York: Oxford University Press, 1996.

Vale, Thomas R. "The Pre-European Landscape of the United States: Pristine or Humanized?" In *Fire, Native Peoples, and the Natural Landscape*. Edited by Thomas R. Vale, 1–40. Washington, DC: Island Press, 2002.

Van den Belt, Henk. "Debating the Precautionary Principle: 'Guilty until Proven Innocent' or 'Innocent until Proven Guilty'?" *Plant Physiology* 132, no. 3 (2003): 1122–1126.

Van den Belt, Henk. "Contesting the Obesity 'Epidemic': Elements of a Counter Discourse." In *Genomics, Obesity and the Struggle over Responsibilities*. Edited by M. Korthals, 39–57. Dordrecht, NL: 2011, Springer.

VandeHaar, Michael J., and Norman St-Pierre. "Major Advances in Nutrition: Relevance to the Sustainability of the Dairy Industry." *Journal of Dairy Science* 89, no. 4 (2006): 1280–1291.

Varner, Gary E. *Personhood, Ethics, and Animal Cognition: Situating Animals in Hare's Two Level Utilitarianism*. New York: Oxford University Press, 2012.

Vileisis, Ann. *Kitchen Literacy: How We Lost the Knowledge of Where Our Food Comes From, and Why We Need to Get It Back.* Washington, DC: Island Press, 2008.

Vogt, Donna U. *Food Biotechnology in the United States: Science Regulation, and Issues.* Washington, DC: Congressional Research Service, 1999. Order Code RL30198.

Wagenvoord, Helen C. "The High Price of Cheap Food." *San Francisco Chronicle Magazine,* May 2, 2004. http://michaelpollan.com/profiles/the-high-price-of-cheap-food-mealpolitik-over-lunch-with-michael-pollan/. Accessed August 25, 2011.

Wambugu, Florence. "Why Africa Needs Agricultural Biotech." *Nature* 400 (1999): 15–16.

Wambugu, Florence. "Why Africa Needs Agricultural Biotech." In *Genetically Modified Foods: Debating Biotechnology.* Edited by Michael Rues and David Castle, 304–308. Amherst, NY: Prometheus Press, 2002.

Weber, Christopher L., and H. Scott Matthews. "Food-Miles and the Relative Climate Impacts of Food Choices in the United States." *Environmental Science and Toxicology* 42 (2008): 3508–3513.

Weirich, Paul, ed. *Labeling Genetically Modified Food: The Philosophical and Legal Debate.* New York: Oxford University Press, 2008.

Welin, Stellan. "Introducing the New Meat. Problems and Prospects." *Etikk i praksis* 1, no. 1 (2013): 24–37.

Wenzel, Siegfried. "The Seven Deadly Sins: Some Problems of Research," *Speculum* 43 (1968): 1–22.

Westra, Laura. "A Transgenic Dinner: Social and Ethical Issues in Biotechnology and Agriculture." *Journal of Social Philosophy* 24 (1993): 213–232.

Wollenberg, Charles. Introduction to *The Harvest Gypsies,* by John Steinbeck. Berkeley, CA: Haydey Press, 1998.

Woolston, Chris. "Republished Paper Draws Fire." *Nature* 511, no. 7508 (2014): 129–129.

World Commission on Environment and Development. *Our Common Future.* New York: Oxford University Press, 1987.

Ye, Xudong et al. "Engineering the Provitamin A (ß-Carotene) Biosynthetic Pathway into (Carotenoid Free) Rice Endosperm." In *Genetically Modified Foods: Debating Biotechnology.* Edited by Michael Ruse and David Castle, 45–51. Amherst, NY: Prometheus Press, 2002.

Yeager, R. F. "Aspects of Gluttony in Chaucer and Gower." *Studies in Philology* 81, No. 1 (Winter 1984): 42–55.

Zerbe, Noah. "Feeding the Famine? American Food Aid and the GMO Debate in Southern Africa." *Food Policy* 29 (2004): 593–608.

Žižek, Slavoj. *First as Tragedy, Then as Farce*. London: Verso, 2009.

Žižek, Slavoj. "Cultural Capitalism." https://www.youtube.com/watch?v=GRvRm19UKdA. Accessed June 19, 2014.

Zwart, Hub. "A Short History of Food Ethics." *Journal of Agricultural and Environmental Ethics* 12, no. 2 (2000): 113–126.

INDEX